Processing and Characterization of Materials ICPCM 2025

7th International Conference on Processing and Characterization of Materials (ICPCM 2025), December 12-13, 2025, NIT Rourkela, India

Editors

Syed Nasimul Alam, Krishna Dutta, Kumud Kant Mehta

Department of Metallurgical and Materials Engineering,
National Institute of Technology Rourkela, Rourkela, Odisha, Pin-769008, India

Peer review statement

All papers published in this volume of "Materials Research Proceedings" have been peer reviewed. The process of peer review was initiated and overseen by the above proceedings editors. All reviews were conducted by expert referees in accordance to Materials Research Forum LLC high standards.

Published under License by **Materials Research Forum LLC**
Millersville, PA 17551, USA

Published as part of the proceedings series
Materials Research Proceedings
Volume 65 (2026)

ISSN 2474-3941 (Print)
ISSN 2474-395X (Online)

ISBN 978-1-64490-412-1 (Print)
ISBN 978-1-64490-413-8 (eBook)

Distributed worldwide by

Materials Research Forum LLC
105 Springdale Lane
Millersville, PA 17551
USA
https://mrforum.com

Manufactured in the United States of America
10 9 8 7 6 5 4 3 2 1

Table of Contents

Preface

7th International Conference on Processing and Characterization of Materials (ICPCM 2025) was held at the Metallurgical and Materials (MM) Engineering Department of the National Institute of Technology Rourkela. The two-day international forum was held on 12–13 December 2025 and was dedicated to materials science and engineering. The conference marks the 7th edition of the ICPCM series, which has been conducted annually since 2011 and is am adaptation of the 15th National Conference on Processing and Characterization of Materials (NCPCM). The event was convened by Prof. Syed Nasimul Alam.

The inaugural session was attended by distinguished guests, including Chief Guest Prof. Lukas Bichler from the University of British Columbia, and Guests of Honour Shri Alok Verma, Director-in-Charge, SAIL-Rourkela Steel Plant, and Prof. Rajesh Prasad from IIT Delhi. Prof. Ashok Kumar Turuk, Dean (Academics), and Prof. Debasis Chaira (HoD, MM Department) were also present on the dais. This year's edition attracted around 60 participants from across the globe, comprising researchers, academicians, and industry professionals.

The event received an overwhelming response from participants around the world. The speaker lineup featured experts from the University of Leoben (Austria), the University of British Columbia (Canada), various IITs and NITs, VIT-AP University, and the Defence Metallurgical Research Laboratory (DMRL), Hyderabad. Industry specialists from TATA Steel, and other organizations were contributing to deliberations on emerging industrial applications of advanced materials. ICPCM 2025 served as a comprehensive platform for presenting research in mineral beneficiation, non-ferrous metallurgy, advanced steels, composites, additive manufacturing, high-entropy alloys, smart materials, corrosion and surface engineering, materials for energy storage, and the use of AI/ML in materials engineering. The two-day event included oral and poster presentations, interactive technical sessions, and the popular metallography contest.

The conference was supported by the Anusandhan National Research Foundation (ANRF), TEXMiN IIT (ISM) Dhanbad (as the Technical Partner), AMETEK Pvt. Ltd, LABINDIA Instruments Pvt. Ltd., Sarvesh Refractories Pvt. Ltd., and Aimil Ltd.

The present volume of the proceedings published by Materials Research Forum LLC compiles the research papers contributed by the participants of the conference. The proceedings consist of selected peer-reviewed papers contributed by participant of ICPCM-2025.

Thanking you!

Dr. Syed Nasimul Alam

Convener,

ICPCM-2025

Committees

Advisory Committee

Prof. Animesh Mandal, Head, School of Minerals, Metallurgical and Materials Engineering, IIT Bhubaneshwar
Prof. Snehnashu Pal, Metallurgical and Materials Engineering Department, IIEST Shibpur, West Bengal
Prof. Sushovan Basak, Mechanical Engineering Department, C. V. Raman Global University, Bhubaneshwar, Odisha
Dr. A. K. Singh, Ex-Scientist G, DMRL, DRDO, Hyderabad
Prof. R. K. Mandal, Ex-Professor, Department of Metallurgical Engineering, Indian Institute of Technology (BHU), Varanasi
Dr. S. Vincent, Department of Mechanical Engineering, BITS Pilani Dubai Campus
Prof. Mohammad Rizwanur Rahman, Department of Metallurgical and Materials Engineering, NITK, Surathkal
Dr. Mayadhar Debata, Scientist, CSIR – Institute of Minerals and Materials Technology (CSIR-IMMT) Bhubaneshwar
Prof. Ashok Kumar Mondal, Department of Metallurgical Engineering, Indian Institute of Technology (BHU), Varanasi
Prof. Suhrit Mula, Department of Metallurgical and Materials Engineering, Indian Institute of Technology Roorkee
Prof. Krishanu Biswas, Department of Material Science and Engineering, Indian Institute of Technology Kanpur
Prof. Tapas Laha, Department of Metallurgical and Materials Engineering, Indian Institute of Technology Kharagpur
Dr. Pratik K. Ray, Department of Metallurgical and Materials Engineering, Indian Institute of Technology Ropar
Prof. Bharat Bhooshan Panigrahi, Material Science and Metallurgical Engineering Department, Indian Institute of Technology Hyderabad

Conference Core Committee

PATRON: Prof. K. Umamaheshwar Rao
Director, NIT Rourkela

CHAIRMAN: Prof. Debasis Chaira
HOD, Metallurgical and Materials Engineering Department, NIT Rourkela

ADVISOR: Prof. Atanu Bhowmick
Metallurgical and Materials Engineering Department, NIT Rourkela

CONVENERS: Prof. Syed Nasimul Alam, Prof. Krishna Dutta and Prof. Kumud Kant Mehta
Metallurgical and Materials Engineering Department, NIT Rourkela

Organizing Committee

• Prof. B. C. Ray • Prof. A. Bhowmick • Prof. A. Basu • Prof. S. K. Sahoo • Prof. A. Mallik
• Prof. N. Yedla • Prof. S. K. Karak • Prof. A. Behera • Prof. A. Patra • Prof. R. K. Prusty
• Prof. S. K. Dinda • Prof. S. Acharya • Prof. A. Sarkar • Prof. Prekshya Nath • Prof. P. G. Ranaware
• Prof. A. A. Kashiwar • Prof. P. Saini • Prof. S. Shukla

All from Metallurgical and Materials Engineering Department, NIT Rourkela

Processing and Characterization of Materials – ICPCM 2025 Materials Research Forum LLC
Materials Research Proceedings 65 (2026) 1-7 https://doi.org/10.21741/9781644904138-1

An overview of iron ore assimilation during particle sintering

B Sai PRAKASH[1,a*], Sriniwas DVARPUDI[1,b]

[1]Research & Development, Agglomeration Research Group, Tata Steel, Jamshedpur, India 831007

[a]saiprakashbph@gmail.com, [b]sriniwas.dvarpudi@tatasteel.com

Keywords: Ore Assimilation, Sintering, Assimilation Theories, Assimilation Quantification

Abstract. Sinter is a precursor input to blast furnace with specific strength, reducibility and chemistry requirements. Assimilation is the primary reaction at the beginning of the sintering process which plays a key role in the obtained sinter properties for blast furnace requirements. The various theories of assimilation in combination with quantification methods were delved into.

1.Introduction

Sinter is a raw precursor input to blast furnace with specific requirements such as fixed variation in chemical composition and size range, with significant tumbler strength and reducibility and minimal reduction degradation index [1].

Sintering is a process that can be beset by many issues that include dynamicity in the nature of raw material, dealing with effects of high alumina (typically >2.5–3.5 wt.% depending on ore type) [2-11], excessive sinter return fines, moisture [12,13], and weather changes affecting sintering, regime shifts, and lastly environmental sustainability[14,15]. The common thread amongst all these issues is understanding the role of assimilation in the sintering of iron ore.

Assimilation is a precursor reaction to the start of the sintering process. Thus, understanding how it influences the reactions through thermodynamic and kinetic studies helps in improving the decisions required for the choice of raw material to be used based on its physicochemical attributes and hence, the techno-economics of the sintering process.

It's important to quantify assimilation to understand how it can be improved further. Characterization studies give an insight into the possible mechanisms at play during assimilation. As the science and theory behind assimilation is developed and with the help of quantification and characterization studies, newer technologies to improve assimilation behavior come to the fore.

Some of the salient features of the assimilation process are that its complementary nature of blending allows us to produce composite properties in the sinter. The process of assimilation indicates a complex interdependence between raw material, process and properties. Lastly, it's a solid-state reaction (often 3-phase) which requires conducive conditions to propagate the assimilation process. The process is explained in greater depth in the following section.

2.Assimilation process:

From studies [16] based on sintering as a process, it essentially is a thermal reaction between coke particles, fluxing agents, and iron ore particles as seen in Figure 1. The entire process proceeds stage-wise with the formation of the initial melt subsequently followed by assimilation of large size particles. It's a contact-based reaction between them preferably with porous ores because the melt penetrates deeper leading to improved assimilation. Section 2.1 will help to understand the consequences of this mechanism on the sintering process.

While Nyembwe's work [16] explains the process of assimilation, a clear demarcation to defining assimilation behavior can help. Also, knowing if assimilation behavior can also include the penetrating ability of the liquid and the ability to coalesce is something that needs to be explored in Nyembwe's work[16].

Materials Research Proceedings 65 (2026) 1-7 https://doi.org/10.21741/9781644904138-1

2.1 Importance of assimilation/relevance to sintering:

As per Debrincat and Loo's work [17], when the iron ore has poor assimilation characteristics, it will melt and thus bond poorly leading to lower sinter strength and lower reducibility of the sinter. On the other hand, a high assimilation characteristic could mean a decrease in the bed permeability [18] in turn affecting the sinter properties such as yield. Hence optimal assimilation characteristics would enable desired sinter properties. Section 3 would enable us to justify various theories for the above experimental observations.

While the author's definitions of assimilation characteristic[17,18] seem useful as an indicator for assimilation it need not necessarily imply that good assimilation leads to good sinter properties because there are other phenomena preceding melt formation, for e.g melt flow, governed by fluidity which can also influence sinter properties.

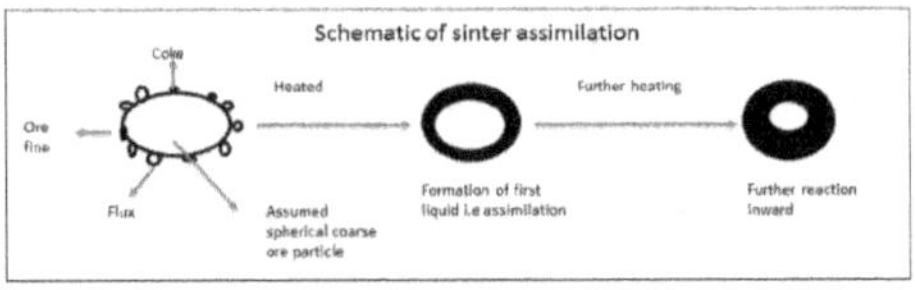

***Figure 1**: Schematic of the assimilation phenomenon taking place during sintering (Redrawn[19])*

3.Theories of Assimilation:

There are numerous approaches to understand, model, and interpret assimilation. Some of them are mentioned below:

3.1. Diffusion model:

A theory has been proposed that the primary melt is formed as the nucleus particles start to assimilate [19]. As the diffusion of calcium ions away from the iron-rich melt is akin to the dissolution of charged species in a melt, a diffusion model has been proposed . The rate of dissolution is expressed as follows:

$$\frac{dm}{dt} = K.A_r.\Delta c^n \quad (1)$$

where m is the net mass of assimilated solid, t is the time, K is a proportionality constant. The rate of dissolution is governed by the solid-melt interfacial area A_r which is perpendicular to the direction of diffusion, temperature, Δc the concentration gradient which is governed by the chemical constitution of the melt. The values of the constant n are governed by the nature of the process. For diffusion, n is unity and is greater than unity for dissolution.

Application:When pisolite ore calcines it becomes very porous as goethite transforms into fine hematite crystallites.Goethitic ores assimilate much more readily than hematite ores because melts are very mobile and easily penetrate into porous ores, which therefore have a very high reaction surface area. Sintering temperature influences both the assimilation reaction rate and the diffusion rate of species. The degree of superheat, which is the difference between the sintering temperature and the liquidus temperature of the primary melt, has been shown to have a significant effect on the viscosity of the melt. Therefore, changes to sintering temperature or primary melt composition will alter assimilation also through changes in melt viscosity.As sintering is carried out close to the liquidus temperature of the melt, relatively small changes in sintering temperature can have a very significant influence on melt fluidity[20,21].

Theory:
Assimilation is governed by mass transfer of ionic species (primarily Ca^{2+}) across the solid–liquid interface, analogous to dissolution processes in melts. The rate of assimilation is controlled by diffusion through a boundary layer formed at the interface. Temperature plays a critical role by influencing both the diffusion coefficient and the viscosity of the melt, thereby affecting mass transfer rates. This framework is particularly relevant in the early stages of assimilation where concentration gradients drive the transfer of species.

Model:
Based on this theoretical framework, the diffusion-controlled assimilation process can be expressed as (1)

3.2 Sinter-block densification:
An analytical approach [22] to understanding the assimilation phenomenon is the usage of polygonal lines to indicate how porous the ores are through the symbolism of lines which are broken to the degree the ore is porous. This approach is used to map how the block strength varies as a function of the volume of melt formed. `The 'block model' helps to study heterogeneity in sinter which is seldom explored because it's chaotic.The author's work[22] is commendable due to it's ability to predict the strength distribution from melt volume,but the model assumes melt volume as the prime contributor towards sinter strength which is not always applicable.

Theory:
Assimilation contributes to sinter strength through progressive densification driven by melt formation and pore filling. As melt is generated and infiltrates the porous structure of the ore bed, it enhances bonding between particles and reduces overall porosity. The degree of densification directly influences the mechanical integrity of the sinter.

Model:
The corresponding model represents the sinter structure as an assembly of polygonal blocks, where strength is expressed as a function of melt volume fraction, pore closure, and connectivity between particles. This structural representation enables visualization of heterogeneity within the sinter matrix and provides a basis for correlating melt formation with strength development.

3.3 Iron-rich depolymerized silica melt :
A similarity has been observed in the way iron oxides melt and solidify for both metallurgical phenomena in the industries as well as igneous rock formation deep under the earth. Industrial sinter is depolymerized silica melts rich in iron and vice-versa for igneous rocks. The similarity lies in thermodynamics taking precedence for both the phenomenon during melting and solidification. As a result, the factors responsible for the assimilation of a nucleus particle in a melt for an industrial sinter play a similar role as that between liquid magma and rock.The author's work[23] is remarkable because the chemical data could be of use to geologists and metallurgists, however an assumption that assimilation is always an endothermic reaction needs to be re-assessed.

Theory:
Assimilation is governed by thermodynamic interactions within a depolymerized silicate melt system enriched with iron oxides, drawing parallels with igneous processes. The structure of the melt is controlled by the degree of polymerization of the silicate network, which in turn affects its fluidity and reactivity. The presence of basic oxides alters this structure and facilitates assimilation.
Model:

The model is based on minimization of Gibbs free energy, where melt behavior is determined by its chemical composition and phase equilibria. The incorporation of oxides such as CaO and FeO leads to depolymerization of the silicate network, thereby increasing melt fluidity and enhancing assimilation. This model provides a thermodynamic basis for understanding melt behavior during sintering.

3.4 Latent heat approach :
The model[24] constitutes the thermodynamics of melt formation and assimilation, ruling out kinetics due to complexity arising from the interaction between temperature, composition and interfacial area. It assumes that the latent heat of solidification and melting are not similar despite the processes looking reversible and hence the concept of melt fraction plays an important role. The formulation used to determine melt-fraction seems to be derived from the lever rule for phases, used in binary phase diagrams.While the model used by the author[24] is very robust in itself due to various sub-models, it tends to overlook the edge effects associated with the sintering phenomenon by virtue of the dimensions of the sinter bed/sinter pot and the geometrical limitations it imposes.

Theory:
Assimilation is governed by energy balance and phase transformation thermodynamics during heating. The formation of melt and its subsequent interaction with solid particles depend on the thermal history and energy input to the system. The transformation from solid to liquid phases plays a crucial role in determining assimilation behavior.

Model:
The model evaluates melt fraction using phase diagram relationships and latent heat considerations. The extent of assimilation is linked to temperature evolution and the energy required for phase transformations. This provides a thermodynamic framework to estimate melt formation and its role in assimilation.

A comparison of all the models is done in Table 1.

Table I: *A comparative study of models used to understand assimilation of sinter*

Name of theory	Governing principle	Advantage	Disadvantage
Diffusion model[19]	Diffusion of calcium ions away from the iron-rich melt is akin to the dissolution of charged species in a melt	Brings predictability to a heterogeneously chaotic process such as sintering. It's also easier to co-relate sintering time with ore charecteristics	Environment surrounding each particle is different and cannot be predicted. So the model is more useful in a controlled environment than industrial applications.

Sinter-block densification[22]	Use of polygonal lines to represent the heterogeneity of ore particles	Helps to map the surface heterogeneity.This is unique because most models assume homogeneity to reduce modelling complications.	Model assumes melt volume as the prime contributor towards sinter strength which is not always applicable.Sinter strength arises from competing factors in raw material and processing parameters.
Iron-rich depolymerized silica melt[23]	Thermodynamics of melting and solidification for igneous melts and melts of industrial sinter	Makes comparative studies easier. Also, certain anomalies in each of the fields can be compared and studied.	Model assumes that assimilation is always endothermic.
Latent heat approach[24]	Thermodynamics of melt and assimilation using latent heat data	Robust model with various sub-models. This enables constant improvement of the sub-models and reduces the complexity of the model.	It overlooks the edge effect.The edge effects become predominant while scaling laboratory pot tests to industrial applications.

The comparative analysis presented in Table I indicates that the existing models capture different facets of the assimilation phenomenon, with each model emphasizing a specific governing mechanism. The diffusion model primarily addresses the kinetics of assimilation through mass transfer considerations, making it suitable for understanding the rate-controlling steps during early-stage reactions. In contrast, the sinter-block densification model focuses on the structural evolution of the sinter matrix, linking melt formation to strength development and heterogeneity within the bed.

Thermodynamic approaches, such as the depolymerized iron-rich melt model and the latent heat-based model, provide insight into melt formation and phase equilibria, highlighting the role of chemical composition and energy balance in governing assimilation. However, these models are often limited in capturing transient and non-equilibrium conditions typical of industrial sintering processes.

A key observation from this comparison is that no single model comprehensively describes assimilation, as the phenomenon is governed by a complex interplay of kinetics, thermodynamics, and structural evolution. Therefore, an integrated approach combining diffusion-driven kinetics, thermodynamic melt formation, and structural densification is essential for a more realistic representation of assimilation behavior in industrial sintering.

While all of the above theories capture assimilation qualitatively, quantification is essential in order to optimize processes.

4.Conclusions

The present review leads to the following key conclusions:
1. Assimilation is a critical early-stage reaction governing sinter quality and process efficiency.
2. High alumina ores adversely affect assimilation through changes in melt properties and fluidity.
3. Diffusion models effectively describe kinetic aspects but are limited under industrial variability.
4. Thermodynamic models provide insight into melt formation but often neglect kinetics and scale effects.
5. No single model fully explains assimilation; integrated thermo-kinetic approaches are required.
6. Future work should focus on combining modeling with advanced characterization techniques for better predictability.

References

[1] Cores, Alejandro, et al.,2013, "Editor's Page: Iron ore sintering. Part 1. Theory and practice of the sintering process." Rev. Avances en Sistemas Informática, 10(1),152-171.

[2] Lu, L., Holmes RJ, and Manuel JR.,2007, "Effects of alumina on sintering performance of hematite iron ores." ISIJ international, 47(3),349-358. https://doi.org/10.2355/isijinternational.47.349

[3] Hino, Mitsutaka, et al.,1999, "Simulation of primary-slag melting behavior in the cohesive zone of a blast furnace, considering the effect of Al 2 O 3, Fe t O, and basicity in the sinter ore." Metallurgical and materials transactions B ,30(4) , 671-683. https://doi.org/10.1007/s11663-999-0028-3

[4] Choudhary, Manoj Kumar, and D. Bhattacharjee.,2008, "Effect of variation of alumina on development of phases during iron ore sintering." ISIJ international ,48(12),1804-1806. https://doi.org/10.2355/isijinternational.48.1804

[5] Sinha, Moni.,2009, "Effect of variation of alumina on the microhardness of iron ore sinter phases." ISIJ international ,49(5) ,719-721. https://doi.org/10.2355/isijinternational.49.719

[6] Cores, Alejandro, et al.,2010, "The influence of different iron ores mixtures composition on the quality of sinter." ISIJ International ,50(8),1089-1098. https://doi.org/10.2355/isijinternational.50.1089

[7] Lu, L.,2015, "Important iron ore characteristics and their impacts on sinter quality-a review." Mining, Metallurgy & Exploration ,32(2) , 88-96. https://doi.org/10.1007/BF03402425

[8] Sinha, Moni, et al.,2015, "Mineralogy of iron ores of different alumina levels from Singhbhum Belt and their implication on sintering process." Journal of Minerals and Materials Characterization and Engineering, 3(03), 180. https://doi.org/10.4236/jmmce.2015.33021

[9] Kalenga, M. K., and Andrie Mariana Garbers-Craig., 2010, "Investigation into how the magnesia, silica, and alumina contents of iron ore sinter influence its mineralogy and properties." Journal of the Southern African Institute of Mining and Metallurgy 11(8), 447-456.

[10] Li, Heping, et al., 2019, "The Mechanism of the Effect of Al2O3 Content on the Liquid Phase Fluidity of Iron Ore Fines." Processes ,7(12),931. https://doi.org/10.3390/pr7120931

[11] O'dea Damien, Ellis Ben. "New Insights into Alumina Types in Iron Ore and Their Effect on Sintering."

[12] Matsumura, Toshihide, et al.,2009, "Effect of moisture absorption behavior on optimal granulation moisture value of sinter raw material." ISIJ international ,49(5),618-624. https://doi.org/10.2355/isijinternational.49.618

[13] Suman, Sekhar, Brijesh K. Giri, and Gour G. Roy.,2013, "Mathematical modelling of iron ore sintering process using genetic algorithm: effect of moisture evaporation and condensation on the temperature profile." Computer Methods in Materials Science 1,141-146. https://doi.org/10.7494/cmms.2013.1.0423

[14] Fernández-González, Daniel, et al., 2017, "Iron ore sintering: Environment, automatic, and control techniques." Mineral Processing and Extractive Metallurgy Review ,38(4), 238-249. https://doi.org/10.1080/08827508.2017.1288118

[15] Lovel, R., K. Vining, and M. Dell'Amico., 2007, "Iron ore sintering with charcoal." Mineral Processing and Extractive Metallurgy ,116(2) , 85-92. https://doi.org/10.1179/174328507X163887

[16] Nyembwe, Mutombo Alainch., 2012, Study of sinter reactions when fine iron ore is replaced with coarse ore, using an infrared furnace and sinter pot tests. Diss. University of Pretoria, https://doi.org/10.1179/1743281213Y.0000000115

[17] Debrincat, D., and Loo CE., 2004, "Effect of iron ore particle assimilation on sinter structure." ISIJ international ,44(8) ,1308-1317. https://doi.org/10.2355/isijinternational.44.1308

[18] Loo, C. E., and J. C. M. Leaney.,2002, "Characterizing the contribution of the high-temperature zone to iron ore sinter bed permeability." Mineral Processing and Extractive Metallurgy ,111.1, 11-17. https://doi.org/10.1179/mpm.2002.111.1.11

[19] Loo, C. E., R. P. Williams, and L. T. Matthews., 1992, "Influence of material properties on high-temperature zone reactions in sintering of iron-ore." Transactions of the institution of mining and metallurgy section c-Mineral processing and extractive metallurgy, 101, C7-C16.

[20] Loo, C. E., 2000, "Changes in heat transfer when sintering porous goethitic iron ores." Mineral Processing and Extractive Metallurgy 109(1) ,11-22. https://doi.org/10.1179/mpm.2000.109.1.11

[21] Loo, C. E., and L. T. Matthews., 1992, "Assimilation of large ore and flux particles in iron ore sintering." Transactions of the Institution of Mining and Metallurgy, Section C(UK), 101.

[22] Nakano, Masanori, and Jun Okazaki.,2011, "Ideal behavior of sinter block densification and relation thereof to yield and strength in iron ore sintering." ISIJ international, 51(9) , 1418-1424. https://doi.org/10.2355/isijinternational.51.1418

[23] Morcos, Riham M., Benjamin G. Ellis, and Alexandra Navrotsky.,2007 "The energetics of hematite dissolution in iron-oxide-rich melts: In situ high-temperature calorimetric studies." American Mineralogist, 92(7),1064-1070. https://doi.org/10.2138/am.2007.2451

[24] Zhou, Hao, et al.,2012 "Numerical modeling of the iron ore sintering process." ISIJ international ,52(9) , 1550-1558. https://doi.org/10.2355/isijinternational.52.1550

Application of agentic AI in composition-property-performance triad for refractories

B Sai PRAKASH[1a*], Shantanu SAHA[1,b], Biswajeet PAL[1,c], Arko CHAKRABORTHY[1,d], Mormee DAS[1,e]

[1]Tata Steel, Jamshedpur-831001, India

[a]saiprakashbph@gmail.com, [b]shantanu.saha@tatasteel.com, [c]biswajeet.pal@tatasteel.com, [d]arko.chakraborthy@tatasteel.com, [e]moromee.dash@tatasteel.com

Keywords: Refractory, Large Language Models, Agentic AI

Abstract:Structure-property-processing remains an integral feature of metallurgical and material science design engineering. Agentic AI uses advanced large language models to access online literature to give estimates of properties based on composition , thus having a deep impact on performance estimation.Initial trials have revealed two such success cases. This has brought down the cost of trials, lab experiment costs and KPIs improvement of parameters such as service temperarure to the tune of 1.2 Cr

Introduction

Thermal conductivity estimation is a huge black-box in the refractory industry due to the expensive and complex nature of high temperature experiments to establish it. An effort was made to use LLMs(Large Language Models) to access online literature to build a relationship between chemistry, grain size and thermal conductivity. In Tata Steel, third parties typically have different refractory grades for certain applications. So this model, helps in assessing the thermal conductivity of those refractories.

Recent literature indicates a strong shift toward artificial intelligence and agentic frameworks for materials design, enabling autonomous decision-making, adaptive optimization, and accelerated discovery pipelines in refractory and high-temperature materials research [1–4]. Machine learning models have been extensively applied for predicting thermo-mechanical properties, wear, corrosion resistance, and phase stability of refractory ceramics and refractory high-entropy alloys, demonstrating significant accuracy improvements over traditional empirical approaches [5–10]. Data-driven and multi-objective optimization strategies, including neural networks, ensemble learning, and physics-informed AI, have further enhanced compositional design and performance targeting under extreme service conditions [11–16]. More recent studies highlight the emergence of agentic and multi-agent AI systems, integrating large language models with materials databases and simulations, marking a transition toward autonomous and self-driving refractory design workflows [17–20]

Framework:

The framework was construed as a combination of computational thermodynamics and 1st principle models to constitute the composition-property-performance framework for refractories in this case. The objective was two-fold: a) Given the composition , what would be the performance of a refractory b) Given the performance what could be the combination of refractories that can be used.(*Fig.1.*) A typical detailed prompt to get a narrowed down answer is shared (*Fig.2.)* and how the input/output response for an answer looks like is shared as well(*Fig.3.)*.

Materials Research Proceedings 65 (2026) 8-14 https://doi.org/10.21741/9781644904138-2

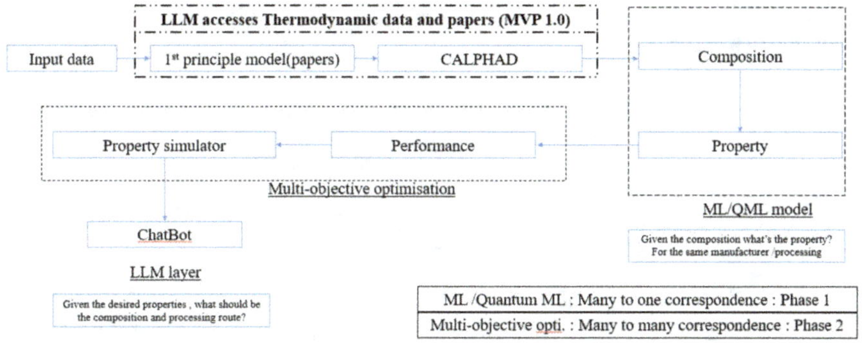

Fig.1: Framework for refractory simulator

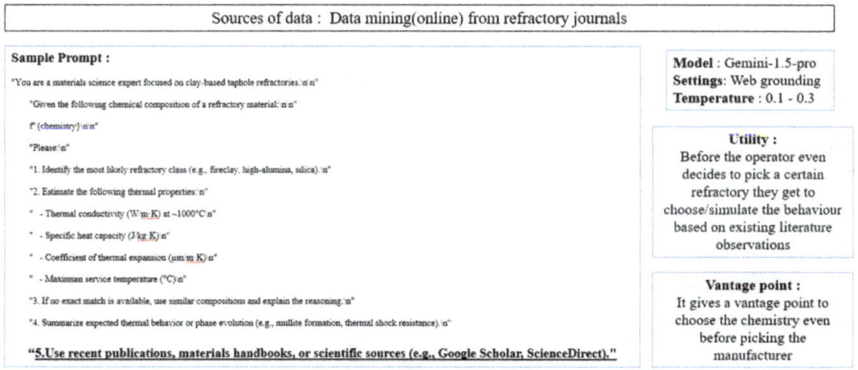

Fig.2: A typical LLM based prompt for user input

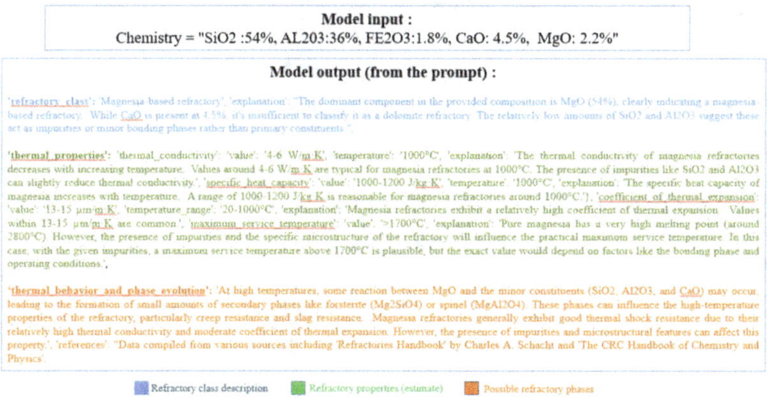

Fig.3: A typical LLM based input/output with relevant literature sources

Findings:

Various trials with respect to web grounding, temperature and model settings were tried and the results are shared (*Fig.4.)*

Without web grinding : Not much change in thermal properties on changing chemistry
On web grounding : Gives an estimate along with the reasons for the same

Temperature (0.1) : Good estimate on specific properties, less detail on phase changes
Temperature (>0.1) : More detail on phase changes , helps to speculate better

GPT-4o-mini vs Gemini 1.5 pro
Gemini 1.5 pro : Gives superior results in terms of better estimates , deeper analysis on phases and references

Conclusions :
Gemini 1.5 pro with web grounding and low temperature for specific properties
Gemini 1.5 pro with web grounding and high temperature for detail on phase changes

Fig.4: Trials and observations on different model settings

In addition to that an analysis on knowledge retrieval bias was done as a qualitative(semi-quantitative) estimate (*Fig.5.*) with regards to knowledge density on certain aspects of refractory literature.

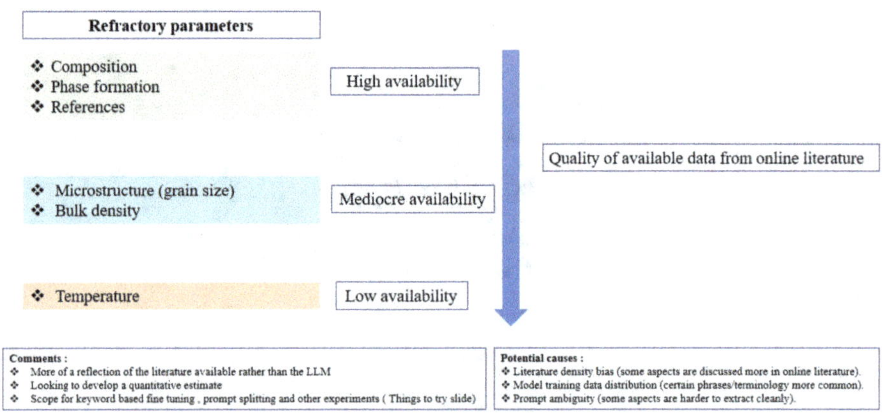

Fig.5: Knowledge retrieval bias : Qualitative estimate of data quality

Materials Research Proceedings 65 (2026) 8-14 https://doi.org/10.21741/9781644904138-2

The study involved multiple controlled trials to understand the behaviour of the agentic AI framework under varying configurations of web retrieval and model parameters. The objective of this phase was not to validate accuracy, but to identify the operational characteristics and sensitivities of the system.

A series of experiments were conducted by varying LLM parameters such as temperature, prompt structure, and retrieval depth (Fig. 4). It was observed that:

Temperature sensitivity:
Lower temperature settings produced more deterministic and consistent outputs, whereas higher temperature values increased variability in predictions. However, higher temperatures occasionally enabled broader exploration of literature, improving contextual richness.

Prompt structure dependency:
Well-defined prompts with explicit constraints (composition ranges, temperature conditions, property definitions) significantly improved the relevance and precision of retrieved information. Generic prompts led to scattered and less reliable outputs.

Retrieval depth influence:
Increasing the number of retrieved sources improved the diversity of information but also introduced noise. An optimal balance between retrieval depth and filtering was necessary to ensure meaningful outputs.

In addition, a qualitative (semi-quantitative) assessment of knowledge retrieval bias was carried out (Fig. 5). It was found that:

A)Refractory systems with high literature density (e.g., commonly used alumina-based systems) yielded more consistent and narrow prediction ranges.

B)Less-documented compositions resulted in wider prediction bands, indicating lower confidence due to sparse data availability.

These findings highlight that the performance of agentic AI systems in materials science is strongly dependent on prompt engineering, parameter tuning, and underlying knowledge distribution.

Results
A retrieval search was done on 2 different bricks . The LLMs predicted range consisted of the actual values (*Fig.6)* .To augment accuracy , in future 2 measures were undertaken a) A preloaded dataset with machine learning model capabilities to predict thermal conductivity b) A 4 tiered web retrieval structure to give accurate readings(*Fig.7.*).

Bricks type	Actual	Predicted
Trough thermal conductivity (TRL Cast -ESS (ML))	4.21	4-5 W/m.K
Steel ladle MgOC bricks (slag zone)	9.6	9-10 W/mK

Fig.6: Actual vs predicted values

Materials Research Proceedings 65 (2026) 8-14 https://doi.org/10.21741/9781644904138-2

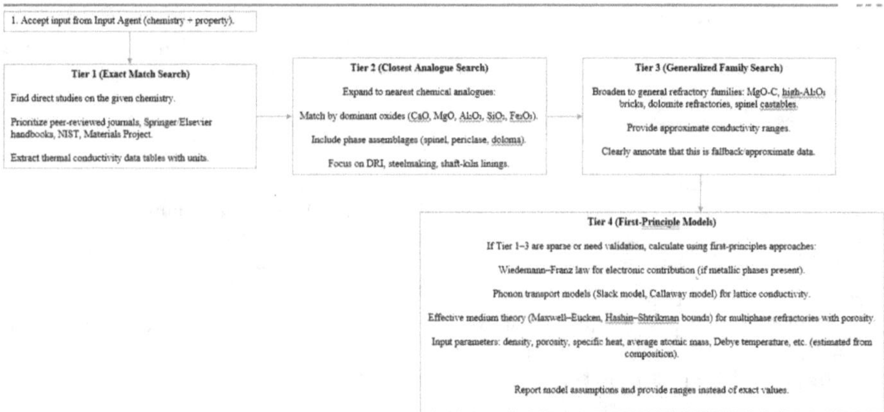

Fig.7: A 4-tiered web retrieval search for accurate thermal conductivity estimate

To evaluate the effectiveness of the proposed agentic AI framework, retrieval-based predictions were carried out on two industrial refractory brick compositions. The objective was to assess whether large language models (LLMs), supported by structured web grounding, could reliably estimate thermal conductivity values within a realistic operating range.

The predicted thermal conductivity values obtained from the LLM-based system were compared against experimentally measured values (*Fig.6*). It was observed that, in both cases, the predicted ranges successfully encompassed the actual measured values. This indicates that the framework is capable of generating **bounded and practically relevant estimates**, even in the absence of explicit experimental input. Such performance is particularly significant given the inherent variability in refractory materials due to differences in grain size distribution, porosity, and phase composition.

A key observation from the trials was that **model configuration parameters**, including temperature settings and retrieval depth, had a notable impact on prediction consistency. Lower temperature settings resulted in more deterministic and reproducible outputs, whereas higher temperatures introduced variability but occasionally improved the breadth of retrieved knowledge. This highlights the need for controlled tuning of LLM parameters depending on whether precision or exploratory insight is prioritized.

Further, the study identified the presence of **knowledge retrieval bias**, as illustrated (*Fig.5)*. Certain refractory systems and compositions exhibited higher representation in available literature, leading to denser and more reliable predictions. Conversely, less-documented compositions resulted in wider prediction ranges, reflecting lower confidence. This semi-quantitative assessment underscores the importance of **data availability and distribution** in influencing LLM-driven materials prediction.

To address these limitations and improve predictive accuracy, two key enhancements were proposed:

1)Integration of a Preloaded Dataset with Machine Learning Models

A curated dataset of refractory compositions and corresponding thermal properties can be used to train supervised machine learning models. This hybrid approach combines the contextual reasoning of LLMs with the numerical precision of data-driven models, thereby reducing uncertainty in predictions.

2)Implementation of a Four-Tier Web Retrieval Architecture
A structured retrieval pipeline was developed (*Fig.7*), consisting of:
A)Primary literature sources
B)Secondary aggregated databases
C)Domain-specific repositories
D)Filtered high-confidence references

This multi-layered approach ensures that retrieved information is both relevant and reliable, improving the robustness of the final prediction.

Overall, the results demonstrate that the proposed agentic AI framework can serve as **a practical decision-support tool** for estimating thermal conductivity in refractory materials. While the current approach provides reliable range-based predictions, ongoing improvements in data integration and retrieval strategies are expected to enhance accuracy further and enable more precise composition-property mapping.

Way forward
Linkage of thermal conductivity to performance occurs through analysis of thickness loss vs temperature , to estimate either of them based on sensors availability leading to improvement in campaign life and other key performance indicators (*Fig.8.)* An estimated benefit of 2.1 cr for steel, trough and DRI kiln refractories has been estimated. The potential benefit on a daily basis is also shared(Table 1.*)*.

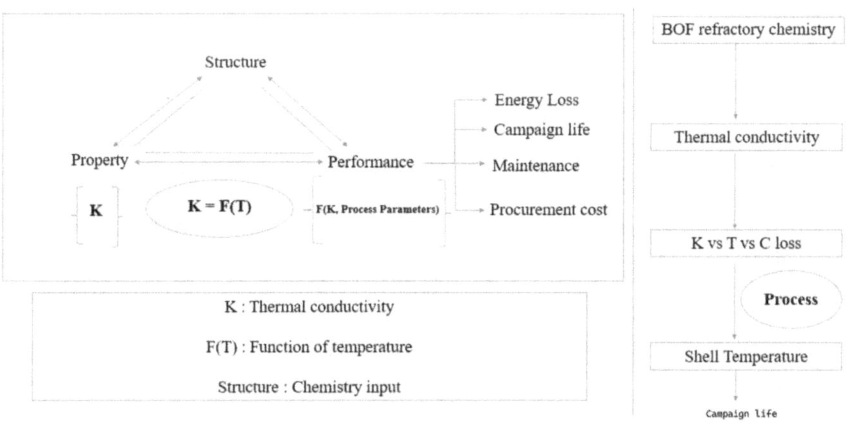

Fig.7: Structure-property-performance linkage

Table1: Estimate of the profit of establishment of the technology

Feature	**What it means**	**Profit to the Company**
Fast & Accurate Thermal Conductivity Prediction	From only compositional input	Saves 2–6 weeks of lab testing per material
Low-Risk Refractory Design & Selection	Suggests best refractory for e.g. specific DRI zone (shaft, bustle, transition, discharge)	Helps prevent furnace shutdowns (₹50 lakhs–₹2 crores/day loss)

References

[1] Chen, Y. et al., Agentic AI and Machine Learning for Accelerated Materials Discovery and Applications, arXiv, 2026.

[2] Gupta, A. et al., Artificial Intelligence in Materials by Design: From Informatics to Agentic Intelligence, Materials Today, 2025.

[3] Nematov, S., Hojamberdiev, M., Machine-Learning-Driven Materials Discovery, arXiv, 2025.

[4] Fu, H. et al., Machine Learning in Materials Science: Performance Prediction, Nanomaterials, 2022.

[5] Chen, Z. et al., ML-Assisted Multi-Element Optimization of Spinel Refractories, Materials, 2025.

[6] Ligezin, P., Pitak, M., Artificial Neural Networks in Ceramic and Refractory Materials, Refractories Journal, 2024. https://doi.org/10.35857/2663-3566.124.13

[7] Seyed Mahmoud, M. et al., ML-Based Design of Refractory Alloys for Thermal Conductivity, Materials, 2023.

[8] Sado, A. et al., Current State of Machine Learning for Refractory Properties, Metals, 2023.

[9] Brunner, S., Deep Learning for Refractory Ceramic Image Evaluation, Springer Nature Community, 2024.

[10] Advanced data mining and AI-based refractory wear prediction, Ceramics International, 2021.

[11] Wu, Z. et al., Machine-Learning-Driven Design of Refractory High-Entropy Alloys, Materials & Design, 2026.

[12] He, F. et al., ML-Assisted Design of RHEAs with Targeted Strength and Ductility, Acta Materialia, 2024.

[13] Xu, Y., Xu, C., ML-Based Multi-Objective Search for Refractory HEAs, Journal of Alloys and Compounds, 2025.

[14] Elkatatny, S., Machine Learning Synergy in High-Entropy Alloys, Journal of Materials Research and Technology, 2024. https://doi.org/10.1016/j.jmrt.2024.10.034

[15] Xie, Z. et al., AI Design of High-Entropy Alloys: Challenges and Opportunities, Metals, 2025. https://doi.org/10.3390/met15091012

[16] Giles, R. et al., Uncertainty Quantification in ML-Based RHEA Design, arXiv, 2023.

[17] Ghafarollahi, A., Buehler, M., Physics-Aware Multi-Agent AI for Alloy Design, MIT, 2025.

[18] Iyengar, A. et al., Composition-Based ML for Phase and Strength Prediction in RHEAs, arXiv, 2025.

[19] Catal, C. et al., Machine Learning-Assisted Alloy Design Systems, Computational Materials Science, 2024.

[20] Qi, J. et al., Integrated Design of Aluminum-Containing Refractory Alloys, arXiv, 2023.

Processing and Characterization of Materials – ICPCM 2025 Materials Research Forum LLC
Materials Research Proceedings 65 (2026) 15-21 https://doi.org/10.21741/9781644904138-3

Enhanced flushing intimation at caster to prevent nozzle clogging using anomaly detection through two algorithms coupled together causing composite signal processing

B Sai PRAKASH[1,a*], Apurba DE[1,b], Amit NAUTIYAL[1,c], Sabyasachi MISHRA[1], Arindam PANDIT[1,d], Moromee DAS[1,e], Yomesh KUMAR[1,f]

[1]Tata Steel, Jamshedpur-831001, India

[a]saiprakashbph@gmail.com, [b]apurbade@tatasteel.com, [c]amitnautiyal@tatasteel.com, [d]arindampandit@tatasteel.com, [e]mormee.das@tatasteel.com, [f]yomeshkmar@tatasteel.com

Keywords: Nozzle Clogging, Flushing, Caster, Anomaly-Detection, Signal-Processing

Abstract. Tata Steel, Meramandali faces severe clogging issues due to which tube change at the nozzle occurs at a frequency of 2 heats. A need for prior intimation to the operator of clogging in IF grade heats (Interstitial free) was required. An anomaly detection system was developed which consisted of exponentially weighted moving average coupled with moving average-standard deviation resulting in 98% accuracy due to composite signal processing arising from dual thresholding causing a benefit of 1.43 Cr due to increase in tube change frequency from 2 heats to 3 heats

1. Introduction

Nozzle clogging in continuous casting is a persistent and critical issue affecting process stability, productivity, and product quality in steel manufacturing. The problem is particularly severe in the casting of interstitial-free (IF) steels, where the interaction between inclusions, steel chemistry, and thermal conditions promotes deposition within the submerged entry nozzle (SEN). Progressive clogging leads to flow restriction, unstable mold conditions, and frequent operational interventions such as nozzle flushing or tube changes.

In industrial practice, clogging is typically inferred indirectly through process signals such as stopper rod position and casting speed. An increase in stopper opening under stable casting conditions often indicates growing flow resistance due to clogging. However, these signals are inherently noisy and influenced by multiple interacting factors, including control system adjustments, superheat variations, and transient disturbances. As a result, distinguishing true clogging behavior from normal process variability remains a significant challenge.

Extensive research has been conducted on SEN clogging mechanisms, including mathematical modeling, inclusion chemistry analysis, and computational fluid dynamics studies. More recently, data-driven and machine learning approaches have been explored for clogging prediction. While these studies provide valuable insights, most of them are **either offline in nature, computationally intensive, or lack interpretability for real-time industrial deployment.** In addition, many approaches do not adequately address the challenge of signal noise and transient fluctuations, leading to unreliable or excessive alarm generation.

Another important limitation in current industrial practice is the lack of a **quantifiable and reliable framework for evaluating flushing effectiveness**. Flushing is widely used as a mitigation strategy; however, its success depends on dynamic process conditions and operator judgment. Without a robust detection system, flushing actions are often reactive rather than proactive, resulting in suboptimal tundish utilization and increased operational cost.

Research Gap:
There is a need for a real-time, robust, and interpretable anomaly detection framework that can:
A)Accurately detect early-stage clogging under noisy industrial conditions
B)Differentiate between transient (micro-level) fluctuations and sustained (macro-level) clogging trends
C)Provide reliable and actionable alarms aligned with operator decision-making
D)Enable quantification and improvement of flushing effectiveness

Objective of the Study:
In this work, a composite anomaly detection model is developed by integrating exponentially weighted moving average (EWMA) with volatility-based adaptive thresholding. The model operates on real-time process signals and incorporates domain-specific constraints such as casting speed stability to improve detection reliability. The primary objective is to provide early and accurate clogging indications, thereby enabling proactive flushing actions and improving tundish performance.

The proposed approach is implemented and validated in an industrial continuous casting environment at Tata Steel, Meramandali. The effectiveness of the model is evaluated in terms of detection accuracy, alarm quality, and operational impact, particularly the increase in heats per tundish from 2 to 3.

2. Data analysis on parameters
SHAP analysis(shapley values) is done for feature importance on parameters used.(*Fig.1*)

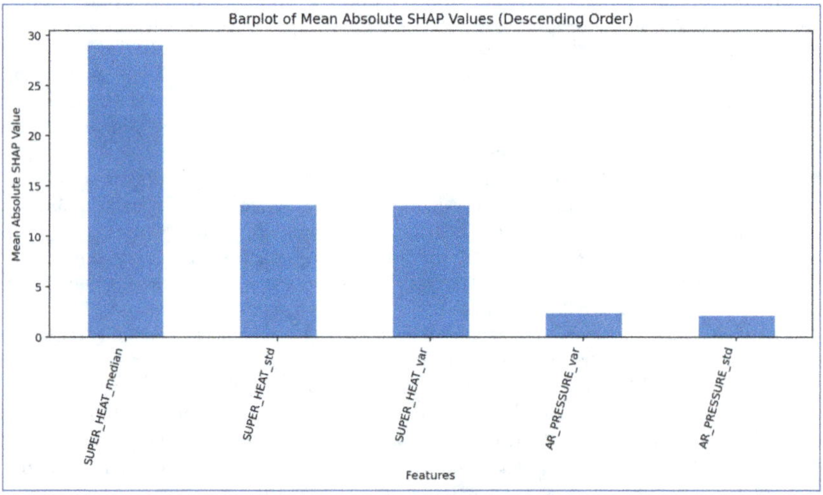

Fig.1 : Feature importance of 2 critical parameters

Superheat (median) is most important .The absolute superheat run at matters more than anything else. The model is keying on the setpoint level, not just its noise. Likely effect is that too low leads to clogging risk and "burst-like" flush events (high peak density but short-lived). Adequate/high leads to smoother flow, fewer peaks. So the action is to revisit the superheat target by grade. Tighten target windows by grade. Std and var refer to standard deviation and variance.

3.Methodology: Formulation of the Clogging Detection Model

3.1 Data Acquisition and Temporal Resolution

Let S(t) denote the stopper rod position and V(t) the casting speed, sampled at discrete time instants tk with a fixed interval $\Delta t = 6$ s. For each heat h, the model operates independently and maintains a rolling buffer of the most recent $N = 1000$ samples. An initial historical backfill of 60 minutes is used to establish sufficient temporal context prior to online evaluation.

3.2 EWMA Smoothing of Stopper Position

The raw stopper position signal is smoothed using an exponentially weighted moving average (EWMA):

$\bar{S}(t_k) = \alpha\, S(t_k) + (1 - \alpha)\, \bar{S}(t_k - 1)$, where $\alpha = 0.1$.

3.3 EWMA Slope as the Primary Trend Indicator

The temporal trend of the smoothed stopper position is evaluated using a first-order discrete difference:

$\Delta\bar{S}(t_k) = \bar{S}(t_k) - \bar{S}(t_{k-1})$.

3.4 Volatility-Based Regime Classification

Stopper position volatility is quantified using a rolling standard deviation over a window of $W = 50$ samples. A stable regime is defined when $\sigma \leq 1.0$, otherwise the regime is classified as volatile.

3.5 Adaptive EWMA Slope Thresholds

Regime-dependent thresholds are applied:

Rstable = 0.005 for stable regime,

Rvolatile = 0.5 for volatile regime.

A rising trend is identified when $\Delta\bar{S}(t_k) > R$.

3.6 Casting Speed Stability Gating

Casting speed stability is evaluated using:

$\Delta V(t_k) \leq 0.002$,

$|V(t_k) - V(t_{k-30})| \leq 0.005$,

$\sigma(v) \leq 0.0015$.

If any condition is violated, trend evaluation is suspended.

3.7 Consecutive Trend Accumulation and Alarm Logic

A counter $C(t_k)$ tracks consecutive valid rising trends. A clogging alarm is generated when $C = 2$, and a flushing alarm when $C = 4$.

Conceptual Interpretation

The clogging detection model is formulated as a deterministic, rule-based framework derived from real-time operational signals in continuous casting. Similar process-based approaches for submerged entry nozzle (SEN) clogging detection have been widely reported in the literature [1–3]. The use of stopper rod position and casting speed as primary indicators is well established in both experimental and industrial studies [4–7].

Stopper position signals are smoothed using an exponentially weighted moving average (EWMA), a technique commonly applied in industrial process monitoring to enhance signal robustness [8,9]. Trend-based evaluation using EWMA slopes has been shown to be effective for early anomaly detection in continuous processes [10,11].

To improve detection reliability, the model incorporates volatility-based regime classification and adaptive thresholds, concepts consistent with earlier studies on SEN clogging under transient and noisy operating conditions [12–14]. Casting speed stability is explicitly enforced to decouple control-induced stopper movements from clogging-related trends, following principles reported in prior casting control and clogging-detection literature [15–17].

Finally, the use of consecutive trend persistence for alarm generation aligns with established practices in industrial alarm management and process diagnostics, ensuring suppression of spurious alarms while retaining early detection capability [18–20].

4. Results

An anomaly detection system was developed which consisted of exponentially weighted moving average coupled with moving average-standard deviation resulting in 98% accuracy due to composite signal processing arising from dual thresholding. (*Fig.2 and Fig.3*). In Figure 2, the model was capturing micro and macro clogging and only macro clogging in Figure 3.

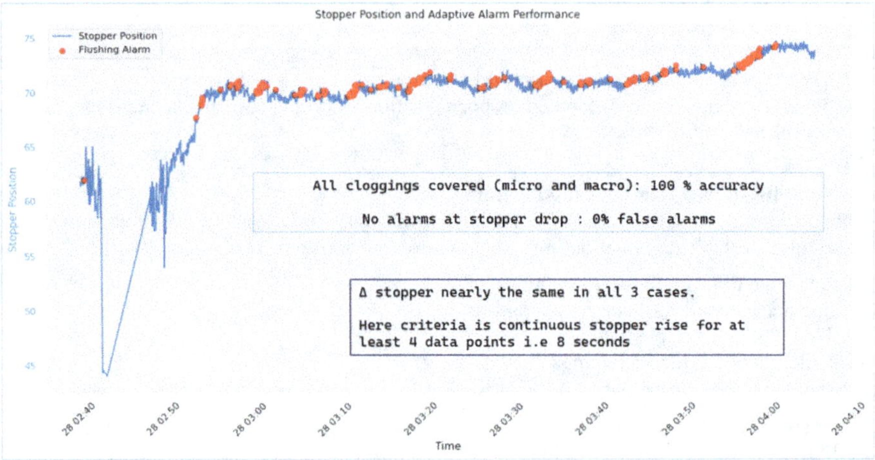

Fig.2. Before scenario : Macro and micro clogging captured (hence many alarms)

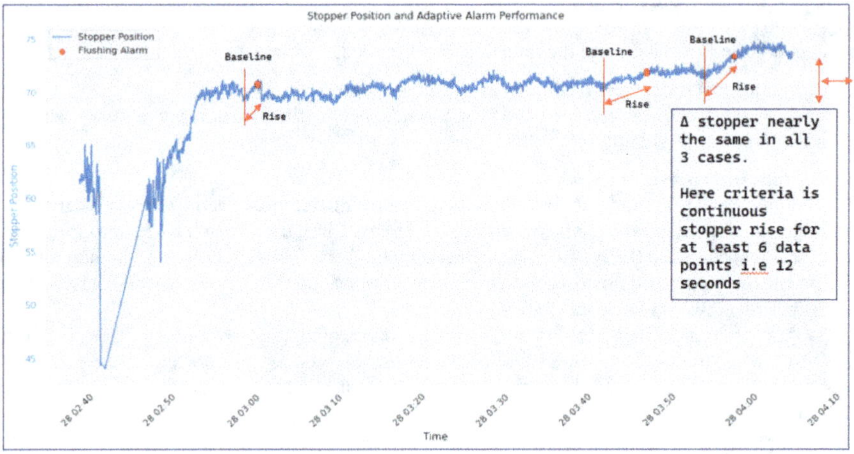

Fig.3. After scenario : Macro clogging captured(only 3 alarms)

Materials Research Proceedings 65 (2026) 15-21 https://doi.org/10.21741/9781644904138-3

The developed anomaly detection system integrates exponentially weighted moving average (EWMA) with volatility-based thresholding to identify clogging events under real-time casting conditions. The model was validated using historical plant data across multiple heats, covering a range of steady-state and transient operating regimes.

The system achieved an overall detection accuracy of **98%**, demonstrating high reliability in identifying clogging events while minimizing false alarms. This performance is attributed to the **composite signal-processing framework**, where EWMA-based trend detection is coupled with volatility-driven adaptive thresholds to enhance robustness under noisy industrial conditions. The model also exhibited strong classification performance, with **precision and recall both exceeding 95%,** indicating minimal false positives and missed detections across varying casting conditions. A key outcome of the proposed approach is its ability to **differentiate between micro-level fluctuations and macro-level clogging trends**, which is critical for practical deployment.

Figure 2 illustrates the **baseline detection behavior**, where both micro and macro variations in stopper position contribute to alarm generation. These micro-level fluctuations are primarily driven by inherent process noise, control system adjustments, and short-term thermal variations. As a result, a large number of alarms are generated, many of which do not correspond to actual clogging progression. This reduces operator confidence and may lead to alarm fatigue.

In contrast, Figure 3 demonstrates the **refined detection output** after incorporating volatility-based regime classification and adaptive thresholding. In this case, only sustained and significant trends associated with macro clogging are detected. The number of alarms is reduced to three, all of which correspond to genuine clogging events requiring operator intervention. This highlights the model's ability to **filter out noise while preserving sensitivity to critical events.**
The improvement can be directly attributed to three key mechanisms:

A) **EWMA smoothing**, which reduces high-frequency noise in stopper position signals (EWMA : Exponentially weighted moving average)

B) **Regime-based threshold adaptation**, which dynamically adjusts sensitivity under stable and volatile conditions

C) **Consecutive trend accumulation**, which ensures that only persistent deviations trigger alarms

Together, these mechanisms enable the system to achieve a balance between **early detection and alarm reliability**, which is essential in continuous casting environments.

From a process perspective, the model effectively captures the underlying physical phenomenon of clogging, where gradual flow restriction leads to sustained increases in stopper position under stable casting speed conditions. By explicitly incorporating casting speed stability as a gating condition, the model avoids false detections arising from control-induced variations.

Operational implementation of the model resulted in a significant improvement in tundish performance. The average tube change frequency increased from **2 heats to 3 heats**, representing a **50% improvement in tundish life**. This directly translates to reduced operational interruptions, improved casting continuity, and enhanced process efficiency.

In addition, the system enables **proactive decision-making** by providing early-stage clogging indications. Operators can initiate flushing actions before severe clogging occurs, thereby preventing unplanned stoppages and maintaining stable casting conditions.

Overall, the proposed framework demonstrates that combining signal-processing techniques with domain-specific constraints can yield a **practical, interpretable, and high-performance solution** for real-time clogging detection in industrial continuous casting processes.

Conclusions

A robust anomaly detection framework for nozzle clogging was developed using EWMA-based trend analysis and adaptive thresholding. The model achieved 98% accuracy with precision and recall exceeding 95%, effectively distinguishing true clogging events from noise. Its implementation enabled proactive flushing, increasing tube life from 2 to 3 heats (≈50% improvement) and delivering measurable operational benefits. The approach provides a practical and interpretable solution for real-time industrial application. A benefit of 1.43 Cr due to increase in tube change frequency from 2 heats to 3 heats resulting in more heats per tundish.

References

[1] Barati, H., Karimi, A., Mathematical modeling of the early stage of clogging of the submerged entry nozzle, Metall. Mater. Trans. A 52 (2021) 4892-4905.

[2] Ikäheimonen, J., Aaltonen, J., Leiviskä, K., Nozzle clogging prediction in continuous casting of steel, IFAC Proc. Vol. 35 (2002) 51-56. https://doi.org/10.3182/20020721-6-ES-1901.01175

[3] Diniz, A.P.M., et al., Use of deep neural networks for clogging detection in the continuous casting process, Expert Syst. Appl. 236 (2024) 121266. https://doi.org/10.1016/j.eswa.2023.121963

[4] Li, Y., et al., Mathematical modeling of transient submerged entry nozzle clogging, Metals 14 (2024) 742. https://doi.org/10.3390/met14070742

[5] Yang, W., et al., Formation and prevention of nozzle clogging during the continuous casting of steels: A review, ISIJ Int. 64 (2024) 1-15. https://doi.org/10.2355/isijinternational.ISIJINT-2023-376

[6] Barati, H., Numerical modeling of clogging in submerged entry nozzle, PhD Thesis, Montanuniversität Leoben, Austria, 2019.

[7] Rackers, K.G., Thomas, B.G., Clogging in continuous casting nozzles, Iron Steel Soc. Conf. Proc. (2004) 1-15.

[8] Thomas, B.G., Tundish nozzle clogging detection and monitoring, Iron Steel Technol. 1 (2003) 39-47.

[9] Kojola, N., et al., Pilot plant study of nozzle clogging mechanisms during continuous casting, Ironmaking Steelmaking 38 (2011) 1-9. https://doi.org/10.1179/030192310X12690127076398

[10] Yu, J., et al., Anti-clogging of submerged entry nozzle through control methods, J. Mater. Process. Technol. 249 (2017) 508-517. https://doi.org/10.1016/j.jmatprotec.2017.06.012

[11] Li, W., et al., Dependence of clogging possibility of the submerged entry nozzle on inclusion composition, Metals 10 (2020) 1205. https://doi.org/10.3390/met10091205

[12] Memarpour, A., A study on submerged entry nozzles respecting clogging, Licentiate Thesis, KTH Royal Institute of Technology, Sweden, 2010.

[13] Rout, B., et al., Development of a nozzle clogging index to assess clogging in continuous casting, Tata Steel Technical Report, India.

[14] Zhang, J., Mathematical model of SEN clogging during continuous casting, Trans. Nonferrous Met. Soc. China 20 (2010) 1534-1540. https://doi.org/10.1016/S1003-6326(09)60334-4

[15] Wang, Y., et al., Influence factors of SEN clogging of rare-earth steel, Steel Res. Int. 94 (2023) 2300078. https://doi.org/10.1002/srin.202300078

[16] Thomas, B.G., et al., Effect of nozzle clogging on mold flow asymmetry, Iron Steel Technol. 9 (2012) 56-64.

[17] Wang, Z., et al., Research progress on application of machine learning in continuous casting, Metals 15 (2025) 1383. https://doi.org/10.3390/met15121383

[18] Thomas, B., Tundish nozzle clogging, Iron Steel Soc. Tech. Note (2001).

[19] Abouelazayem, S., et al., Switched MPC based on clogging detection in continuous casting, IFAC-PapersOnLine 53 (2020) 11539-11544. https://doi.org/10.1016/j.ifacol.2020.12.589

[20] Barati, H., et al., Role of solidification in submerged entry nozzle clogging during continuous casting of steel, Metall. Mater. Trans. B 52 (2021) 3271-3284. https://doi.org/10.1007/s11663-021-02336-x

Materials Research Proceedings 65 (2026) 22-29 https://doi.org/10.21741/9781644904138-4

Tuning the mechanical properties of composite material

S.K. PRADHAN[1,a*], Ananya SRIVASTAVA[1,b], Shreyanshi MALLIK[1,c], P. KOUR[2,d]

[1]Dept. of Mech. Engg., BIT Mesra, Patna campus, Patna, India

[2]Dept. of Physics, BIT Mesra, Patna campus, Patna, India

[a]sudipta.kishore@bitmesra.ac.in, [b]ananyasri205@gmail.com, [c]shreyanshimallik@gmail.com, [d]paramjit.kour@bitmesra.ac.in

Keywords: PVDF, SEM, UTM

Abstract. The study investigated the material properties of White Cement (WC) composites reinforced with Coconut Fibre (CF) (5%, 10%, 15%, 20% by weight) and Polyvinylidene fluoride (PVDF) (5%, 10% by weight) for prospective application in acoustic and transducer insulation. The composites were prepared by direct mixing of white cement, water, and coconut fibre (stirred, molded, and cured for 24 hours) and by dissolving PVDF in Dimethylformamide (DMF) by means of a magnetic stirrer before mixing with white cement. Coconut fibre enhances the toughness and crack resistance of cement. Studies indicate that increased coconut fibre concentration improves acoustic absorption. However, high concentrations (>10%) may lead to reduced workability and lower mechanical strength due to fiber agglomeration and weak inter-transition zones. PVDF acts as a piezoelectric component, which is valuable for transducers. Integration PVDF with white cement allows for the growth of smart, flexible piezo-composites with fine acoustic impedance matching. The stress and strain graphs by UTM (Universal testing machine) likely show enhanced post-cracking ductility and toughness with coconut fiber, whereas PVDF composites show improved flexibility. The addition of fiber generally increases the toughness but may slightly decrease the overall surface hardness compared to pure cement. SEM (scanning electron microscope) analysis would typically disclose the bonding between the coconut fiber and cement matrix (showing good adhesion or voids) and the distribution of PVDF in the matrix.

Introduction

Research on the mechanical and piezoelectric characteristics of composites involving Polyvinylidene Fluoride (PVDF), white cement, and natural reinforcements like coconut fibre aims to enhance the strength and performance of active transducers. PVDF, is having high piezoelectric coefficients, low dielectric loss, and flexibility. PVDF with cement form 0-3 type composite. Adding coconut fibre to cement-based composites improves their mechanical strength, specifically enhancing toughness, ductility, and impact resistance. It gives better performance in harsh conditions.

The intent of current work is to study about mechanical characteristic of piezoelectric material viz. Polyvinylidene Fluoride (PVDF) composite with white cement. Reinforcement restrict post ductility of composite[1]. Reinforcement restricts the post-yield ductility of composites primarily through mechanisms that enhance strength and stiffness at the expense of deformation capacity such as matrix constraint, high load transfer to brittle reinforcements, and strain localization. It has been studied mechanical characteristic of white cement and coconut fibre composite with the intention to seek out the possibilities of strength improvement of piezoelectric material due to addition of coconut fibre. Coconut fibre with different proportion is being mixed [2].Fibre retard development of cracks in composite[3]. So that it can be used in harsh and hostile conditions. Impact resistance is being increased[4].There are several transduction phenomenon which are being used to make a transducer but it is always recommended to use a transducers which are active in nature since these transducers do not need any external power source for their working. Impact resistance increases due to addition of coconut coir fibre with cement [5,6].Coconut coir is environmentally free and inexpensive [7].Mechanical strength varies with size and shape of fibre[8].Fly ash improve the strength of cement[9,10].Coconut fibre is strong, flexible and durable strands[11,12]. Coconut and jute fibre improves impact resistance of cement [13]. One of the potential alternatives is the incorporation of natural fibers into the cement matrix, replacing synthetic fibers and steel. These fibers can reduce production costs by enhancing the tenacity, lightness, mechanical strength, impact resistance, and biodegradability of cement-based composites. The matrix distributes part of the load to the fibers during loading before any cracks appear [14].Due to above mentioned reason it has been focused on piezoelectric material because piezoelectricity is an active phenomenon. Due to ability to withstand curing process of cement, PVDF-Cement composite is being used to create smart ,self-powered sensors embedded directly into structure. The coconut coir fibre-cement composite is being used as ecofriendly construction material, which improve crack resistance in cement. It is mainly used in acoustic insulation and low cost green building construction.

Methodology

It has been concerned about mechanical characteristics of composite of Polyvinylidene fluoride and white cement and composite of coconut fibre and white cement. So, to know about mechanical characteristics of these material sample has been made in cylindrical shape and tested these sample on Universal testing machine and Rockwell hardness tester. Sample of coconut fibre and white cement has been prepared by mixing the coconut fibre and white cement properly in a small container then 40% water added. After adding water the sample has been mixed with the help of a stirrer, after mixing the sample poured in the mould for aging. After 24 hours the sample has been taken from the mould and tested these sample on Universal testing machine and Rockwell hardness tester. Sample of PVDF and white cement composite prepared with the help of magnetic stirrer and ultrasonic vibrator .

Three gram Polyvinylidene fluoride were dissolved in 40ml Dimethylformamide (DMF) solvent with the help of ultrasonic vibrator and magnetic stirrer. The whole mixing process took 47 minute on ultrasonic vibrator and 30 minute on magnetic stirrer. After dissolving PVDF in DMF solvent it has been added this solution to 27g white cement and stirred it properly for several minute to mix it properly, then poured the material into moulds to make two sample of similar kind. Then it was left the sample for 72 hours for binding and forming a solid composite of PVDF and White cement. Another 1.5g Polyvinylidene fluoride were dissolved in 40ml Dimethylformamide (DMF) solvent with the help of same equipment. This time the whole mixing process took 50 minute on ultrasonic vibrator and 12 minute on magnetic stirrer. After dissolving PVDF in DMF it use this solution in 28.5g white cement and mixed it properly with the help of stirrer, then poured the sample in mould and left it for 72 hours. For the purpose of evaluation of strength it is made a sample of pure white cement. For 5% sample it took 2.5g coconut fibre and 47.5g white cement and mixed it in dry state for better dispersion of coconut fibre in white cement. Then it has been added 20ml of water and again stirred the sample properly after mixing the sample properly poured the material into the mould and left for 22 hours. After 22 hours samples were ready for testing. It was repeated the process for 10%, 15% and 20% samples.

Results and Discussion

In figure1. with addition of PVDF compressive strength of composite were reduced but brittleness were completely eliminated. For detection of compressive strength universal testing machine (UTM) is being used. In the result 5% PVDF composite has least compressive strength which means there is effect of concentration of solution of PVDF and DMF. Curves of 5% and 10% samples are most near to elastic plastic with strain hardening. As it can be observed that stress increased very rapidly with slight increment of strain. It can be concluded from result that brittleness of material is removed because there is no fracture in sample having PVDF content.

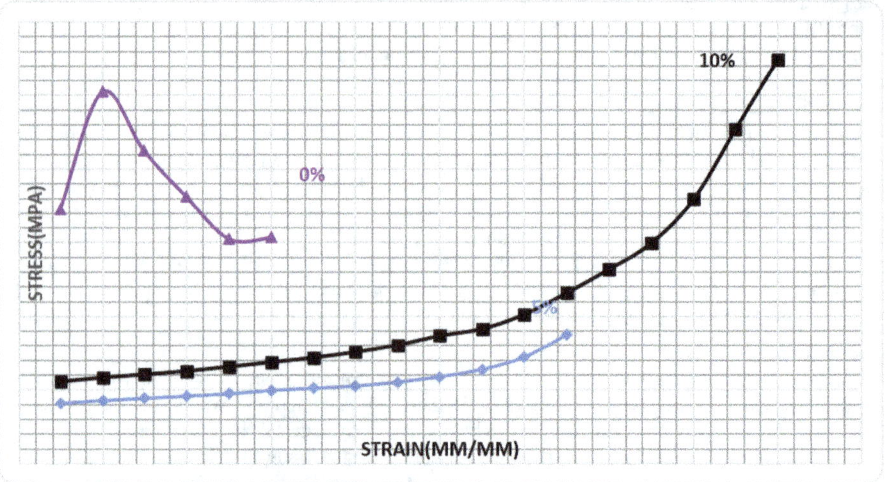

Figure 1.Comparison among 0%, 5% and 10% PVDF in white cement composite

In figure2. The addition of coir fibre initially reduces the compressive strength of white cement, with the impact being more harsh at lower value. At 20% the compressive strength increases again, signifying that high amounts of coir fibre have a less harmful effect on strength compared to small amounts. The composite performance shifts from brittle to ductile as the coir fibre content increases. While 0%, 5%, and 10% samples show a distinct fracture point (brittle failure), the 15% and 20% samples lack a specific fracture point, indicating that increased fiber content reduces brittleness and can almost eliminate it. The stress-strain curves for the 15% and 20% samples are closer to an elasto-plastic material with strain hardening. The sudden increase in stress after 1.2mm/mm of strain in these samples confirms that the addition of higher amounts of coir fibre induces strain hardening. With 10% coir clustering of fibers tends to occur, resulting in a weak zone and increased porosity.With15% coir high concentrations hinder proper compaction of the cement mix. With 20% coir despite higher strength than intermediate samples, this high dosage leads to increased porosity.

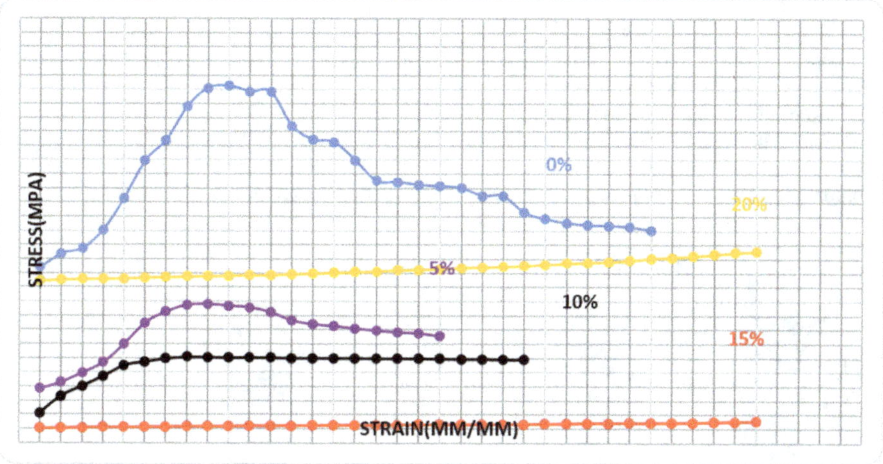

Figure 2. comparison of 0%, 5%, 10%, 15% and 20% coir Fibre and White cement composites

In figure3.hardness of pure white cement is more than composite of PVDF and White cement ,which means addition of PVDF in White cement reduces its hardness. Hardness is being measured using Rockwell hardness tester. With 10% PVDF porosity decreases with increase in hardness. It has been found that hardness is less for 10% PVDF as compared to pure white cement. Hardness test is crucial material ability to withstand structural loads without fracture during its use as sensor.

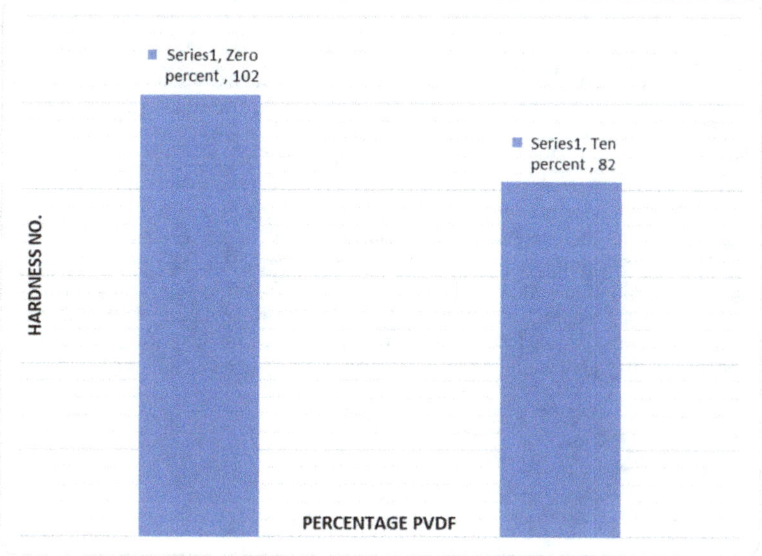

Figure 3. Comparison of hardness of PVDF and White cement composite

From figure4. Based on the analysis of coir fiber-reinforced cement composites, the addition of natural fibers generally reduces the hardness of the matrix due to increased porosity. The 10% sample exhibits the least hardness, because the high volume of fiber requires additional water to maintain workability, resulting in increased voids (capillary pores) upon curing. Excess water and increased fiber content avert proper densification of the matrix, creating more voids, which appreciably lowers the compressive and surface hardness properties. While coir fibers can improve impact strength and ductility, they frequently decrease compressive strength and hardness compared to pure cement paste, particularly when the fiber content is high (e.g., above 1.5–2.5%.

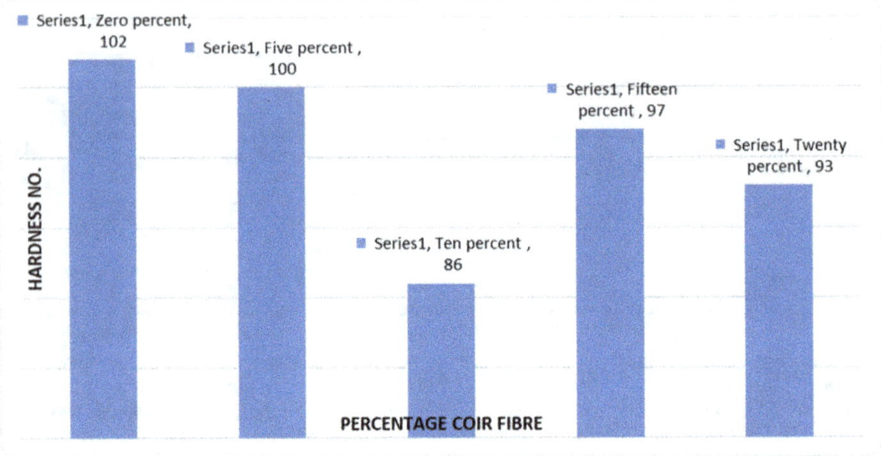

Figure 4. Comparison of hardness of Coir fibre and White cement composites

Figure5.(a) shows SEM for pure PVDF sample[15-17].SEM is used to measure morphology for fracture surface of composite. In figure5.(c) 10% sample grain and grain boundaries are less visible and shows small grain ,grain boundaries. In figure 5(b) 5% sample grain and grain boundaries are clearly visible i.e. prominent grain and grain boundaries. Cement grains are settled down in PVDF due to high density of cement. The cement grains slow down within the PVDF matrix due to the high density of the cement compared to the polymer matrix. This interpretation implies that increasing the cement content (from 5% to 10%) changes the fracture morphology from a distinct grain structure to a more indistinct/packed structure due to the filler distribution. Scanning Electron Microscopy (SEM) is essential for PVDF-reinforced white cement composites to visualize filler dispersion, interfacial bonding, and pore filling, directly impacting piezoelectric performance and acoustic impedance. It validates microstructural integrity, crucial for optimizing transducer sensitivity and detecting voids that degrade acoustic applications.

Figure 5.(a). SEM result of pure PVDF

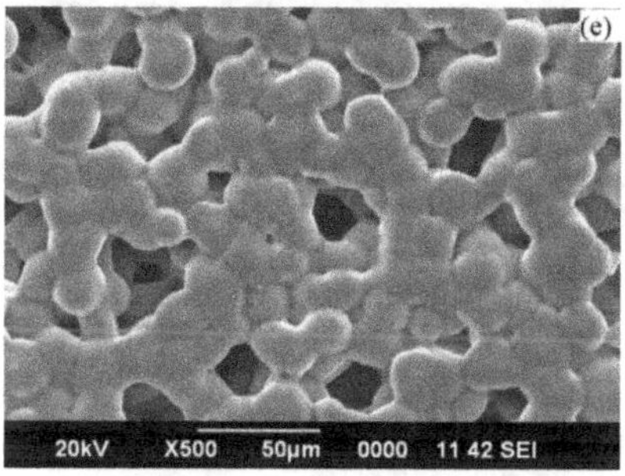

Figure 5.(b).SEM result of pure 5% PVDF in white cement

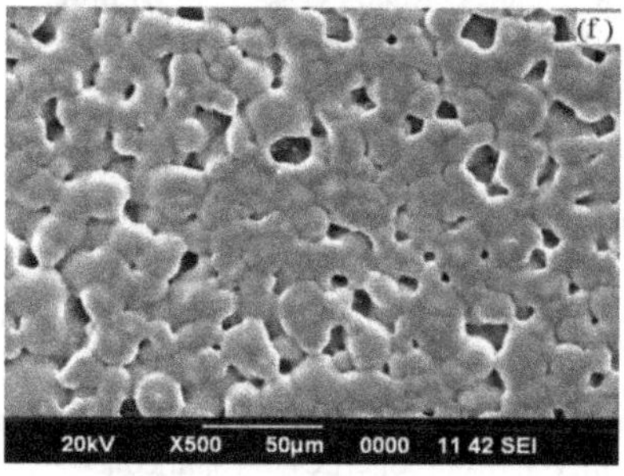

Figure 5.(c).SEM result of pure 10% PVDF in white cement

Conclusions

Based on the study described, the exploration of Polyvinylidene fluoride (PVDF)/white cement and coconut fiber/white cement composites for transducers and structural health monitoring (SHM) sensors indicates a transaction between sustainability and mechanical strength. Both composite types (PVDF/white cement and coconut fiber/white cement) showed a reduction in compressive strength compared to pure white cement. Contrary to general strength reduction, within the coconut fiber and white cement composites, higher amounts of coconut fiber resulted in higher compressive strength compared to samples with lower fiber amounts. The hardness of all composites was reduced upon the addition of either PVDF or coconut fiber. Scanning Electron Microscopy (SEM) was used to analyze the bond quality and distribution of PVDF fibers within the composite matrix. Coconut fiber introduces environmental sustainability to the sensors. While strength decreases, the inclusion of organic fibers like coconut can improve toughness, whereas PVDF is known for its excellent piezoelectric properties in sensor applications.

References

[1]AM. Sunny , MF. Shahriar , M.Mahmud , Finite element parametric study of bearing strength of unconfined cold-formed steel bolted connection, Adv Eng Res. 2025. https://doi.org/10.2991/978-94-6463-672-7_12

[2]P. Mahakavi ,R. Chithra . Impact resistance, microstructures and digital image processing on self-compacting concrete with hooked end and crimped steel fiber. Constr Build Mater. 220:651–66.2019 10.1016/j.conbuildmat.2019.06.001

[3] MS.Islam , SJ.Ahmed , Influence of jute fiber on concrete properties, Constr Build Mater. 189:768–76.2018. https://doi.org/10.1016/j.conbuildmat.2018.09.048

[4]S. Ozen et al, Effect of graphene oxide-coated jute fiber on mechanical and durability properties of concrete mixtures.,Constr Build Mater. 448:138225.,2024. https://doi.org/10.1016/j.conbuildmat.2024.138225

[5] FRB.Martinelli , FRC.Ribeiro , MT.Marvila , SN.Monteiro , Filho FdaCG, Azevedo ARGde. A review of the use of coconut fiber in cement composites. Polymers. 15(5):1309.,2023 https://doi.org/10.3390/polym15051309

[6] KP.Guruswamy , S.Thambiannan , A.Anthonysamy , K.Jalgaonkar ,AS. Dukare , R.Pandiselvam , et al. Coir fibre-reinforced concrete for enhanced compressive strength and sustainability in construction applications. Heliyon. 10(21),2024. https://doi.org/10.1016/j.heliyon.2024.e39773

[7] MB.Khan , N.Shafiq , A.Waqar , D.Radu , C.Cismaş , M.Imran , et al. Effects of jute fiber on fresh and hardened characteristics of concrete with environmental assessment. Buildings. 13(7):1691.2023. https://doi.org/10.3390/buildings13071691

[8] Abu-Bakr.Masood, HF. Mahmood, Influence of using different types of fibers on the mechanical properties of roller compacted concrete, A review. The Journal of the University of Duhok, 25(2), 458–472,2022. https://doi.org/10.26682/sjuod.2022.25.2.42

[9] SA.Ahmad , AJ.Nimat , HF.Mahmood . Effect of different types and ratio of fibers on the tensile strength of reinforced foam concrete, a review. Emerg Technol Eng J.1(2):72–83.,2024 https://doi.org/10.53898/etej2024125

[10] SA.Ahmad ,HF. Mahmood , Omar Fqi K., An overview of the impact of fly ash and polypropylene fiber on the mechanical properties of foam concrete. J Stud Civil Eng. 1(2):75–86.2024. https://doi.org/10.53898/jsce2024125

[11] AP.Martins , RA.Sanches . Assessment of coconut fibers for textile applications. Matéria (Rio de Janeiro). 2019. https://doi.org/10.1590/s1517-707620190003.0743

[12] J.Ahmad , A.Majdi , A.Al-Fakih , AF.Deifalla , F.Althoey , El Ouni MH, et al. Mechanical and durability performance of coconut fiber reinforced concrete: a state-of-the-art review. Materials. 15(10):3601.2022. https://doi.org/10.3390/ma15103601

[13] W.Ahmad , SH.Farooq , M.Usman , M.Khan , A.Ahmad , F.Aslam , et al. Effect of coconut fiber length and content on properties of high strength concrete. Materials. 13(5):1075. 2020. https://doi.org/10.3390/ma13051075

[14]S.Teng , V.Afroughsabet .,C.P. Ostertag . Flexural behavior and durability properties of high performance hybrid-fiber-reinforced concrete. Constr. Build. Mater. 182:504–515.,2018. https://doi.org/10.1016/j.conbuildmat.2018.06.158

[15] SK. Pradhan, P .Kour, Amit Kumar, Rabichandra Pandey, Pawan Kumar, Manoranjan Kar, AN. Sinha. Effect of ferroelectric filler nanoarchitectonics on the electrical and mechanical properties of the nanocomposite thick films of polyvinylidene fluoride and lanthanum-doped Lead zirconate titanate 0-3 connectivity,Applied Physics A,129(11),805,2023. https://doi.org/10.1007/s00339-023-07058-

[16] SK Pradhan, A .Kumar, P. Kour, R. Pandey, P. Kumar, M. Kar, AN. Sinha. Piezoelectric and mechanical properties of PVDF-PZT composite,Ferroelectrics, 558 (1), 59-66,2020. https://doi.org/10.1080/00150193.2020.1735889

[17]H.Zhang,Y.Zhu,L.Li. Fabrication of PVDF/graphene composites with enhanced b phase via conventional melt processing assisted by solid state shear milling technology,Royal soc.of Chemisty,10,3391,2020. https://doi.org/10.1039/c9ra09459h

Surface modification of hemp for improving mechanical performance of sustainable bio-material composites

Devalingam Santhosh Kumar[1,2,a], Rohith Gandi[1,b], Vansh Vandhe[3,c], Tanmay Bhattacharyya[4,d], Rajesh Kumar Prusty[5]

[1]FRP Composites Lab, Department of Metallurgical and Materials Engineering, National Institute of Technology Rourkela, India – 769008

[2]Department of Mechanical Engineering, MVGRCE Vizianagaram, Andhrapradesh, India

[3]Department of Mechanical Engineering, MITS, Gwalior, Madhya Pradesh, India

[4]New Materials Business, Tata Steel Limited, Khopoli, Mumbai India

[5]Center for Nanomaterials, National Institute of Technology, Rourkela, 769008, India

[a]922mm5001@nitrkl.ac.in, [b]rohithnitrkl@gmail.com, [c]prustyr@nitrkl.ac.in, [d]nitrkl.rajeshprusty@gmail.com

Keywords: Hemp, Natural Fibre, Surface Treatment, Flexural, LSS

Abstract. This study focuses on the surface modification of hemp fibres through alkali-based chemical treatment using sodium hydroxide (NaOH), potassium hydroxide (KOH), and sodium bicarbonate ($NaHCO_3$). The influence of chemical treatments on the flexural, short beam shear, and thermo-mechanical behaviour of hemp fibre-reinforced epoxy composites was systematically investigated. The chemical treatment improved the interfacial adhesion between the fibre and the epoxy matrix, as later underscored by the enhanced flexural and short-beam shear performance. The flexural strength of composites fabricated using NaOH-treated, KOH-treated, and $NaHCO_3$-treated hemp fibres is higher by ~4%, ~15%, and ~26%, respectively, than untreated hemp/epoxy composites. Similarly, the flexural modulus of composites fabricated using NaOH-treated, KOH-treated, and $NaHCO_3$-treated hemp fibres increased by 16%, 18%, and ~32%, respectively. When compared to untreated hemp/epoxy, the interlaminar shear strength of composites fabricated using NaOH-treated, KOH-treated, and $NaHCO_3$-treated hemp fibres was higher by ~12%, ~18%, and ~29%, respectively. Topography of the hemp fibre observed during scanning electron microscopy revealed rougher hemp fibres surfaces free of impurities. The FTIR analysis further confirmed the removal of hemi-cellulose.

Introduction

Lignocellulosic fibres have gained popularity in recent decades as reinforcements in both thermoplastics and thermosets due to their promising properties, such as low density, wide availability, biodegradability, recyclability, and high polymer wettability. Furthermore, in contrast to synthetic fibres like glass, carbon, and Kevlar, these fibres are renewable and have a CO_2-neutral life cycle. Ligno-cellulosic fibres can be collected from several parts of the plant, but they are most obtained from the leaves (date palm, banana, sisal, etc.) or the bast. Bast fibres, particularly flax, hemp, and kenaf, are regarded as the most promising due to their characteristics and availability. They have similar morphologies and can have similar functions in the composite. Hemp fibre primarily consists of cellulose, hemicellulose, lignin, and pectin. The physical characteristics of the fibre mainly depend on these contents. Cellulose is the most rigid and strongest organic component of the hemp fibre [1]. Hemp fibre's high tensile strength, stiffness, high aspect ratio (length/diameter ratio), and lower density make it ideal for reinforcement in composite structures. However, similar to other natural fibres, they have drawbacks such as uneven and rough surfaces, variable properties, and low resistance to water absorption and degradation [2]. As the matrix is hydrophobic, the hydrophilic nature of hemp fibre, due to the presence of wax and pectin, limit an efficient reaction with the epoxy thermoset, preventing the reinforcement and matrix from bonding more effectively[3]. As a result of this weak interface, the mechanical characteristics of hemp/epoxy composites were reduced [4], [5]. So, the surface of hemp fibres is modified through various chemical treatments like alkali, silanization, benzoylation, and acetylation. These treatments decrease the fibres' affinity for water and lower the cellulose crystalline index, which enhances their compatibility with the polymer matrix, mechanical properties, and dimensional stability. Alkaline treatment (or mercerization) is one of the most commonly used low-cost alternatives compared with other treatments, such as acetylation and silanization. This treatment transforms hemp fibre morphology through a controlled chemical process where solution breaks the fibre bundles into smaller fibrils by dissolving hemicellulose and lignin. Several factors influence the effectiveness of alkaline treatment, with the type and concentration of the alkaline chemical being the most critical.

Materials

The hemp fabrics and epoxy resin were used respectively as the reinforcing and matrix components. Plain-woven hemp fabrics were sourced from Go Green Products, Chennai. The epoxy resin, Diglycidyl ether of bisphenol-A, was procured from Atul Industries, Gujarat, along with the curing agent, Triethylene tetramine. To pre-treat the hemp fabrics' surface, Sodium Hydroxide (NaOH), Potassium Hydroxide, and Sodium bicarbonate ($NaHCO_3$) pellets, of molarity 40M, were sourced from "Merck Life Science Pvt Ltd, Mumbai.

Surface Treatment of Hemp fabrics

Surface treatment of hemp fabrics was done using a 5% (w/w) solution of NaOH, where water and ethanol were mixed in an 80:20 volumetric proportion. The solution was agitated at room temperature for 1 hour. Later, hemp fibre fabrics were immersed in a previously prepared 5% NaOH solution. After immersion of the hemp fabrics in NaOH solution, they were left undisturbed for 3 hours. The soaked hemp fabrics were washed several times with tap water to remove any residual NaOH groups on the fibre surfaces. Drying of the treated samples was done first by keeping them in air at room temperature for a total of 12 hours. Later, they were dried in a hot-air oven at 100°C for a total of 6 hours. After complete evaporation of water content in samples, they were packaged in plastic bags that are afloat in nature. This was done to avoid absorbing moisture present in atmospheric conditions.

Fig 1. Schematic illustrating the alkaline treatment process of hemp fibres.

Materials Research Proceedings 65 (2026) 30-34 https://doi.org/10.21741/9781644904138-5

The same procedure was repeated with other chemicals, KOH and $NaHCO_3$. Figure 1 illustrates the alkaline treatment process followed for hemp fibres. The NFRP composites fabricated using untreated, NaOH-treated, KOH-treated, and NaHCO3-treated hemp fibres were mentioned as UTHE, NaOH HE, KOH HE, $NaHCO_3$ HE, respectively.

Experimental characterization

Using universal testing machine (Make& model: Instron 5967), the flexural test were conducted according to the ASTM standard D7264, and short beam shear (SBS) tests were conducted by adhering to the ASTM standard D2344 for all treated and untreated hemp/epoxy composites. A Fourier-transform infrared spectrometer (Make & model: Shimadzu IR P-21) was employed to examine the chemical changes in hemp fibre after chemical treatment. Topography and fractography were conducted for both treated, untreated fibres, and composites respectively using a field emission scanning electron microscope (Make & model: FEI Novanano SEM 450).

Results and discussion

Figure 2 depicts the peak intensity and corresponding groups of infrared spectra from hemp fibres treated with NaOH, KOH, and $NaHCO_3$. The O-H stretching band assigned to 3325 cm^{-1} for untreated fibres showed a slight shift and a large difference in intensity for treated fibres. This demonstrates a reduction in hydrogen bonding in cellulose hydroxyl groups, which results in a decrease in hydrophilic behavior. The absence or drop in intensity of the primary hemicellulose (1724 cm^{-1}) showed a decrease in non-cellulosic content.

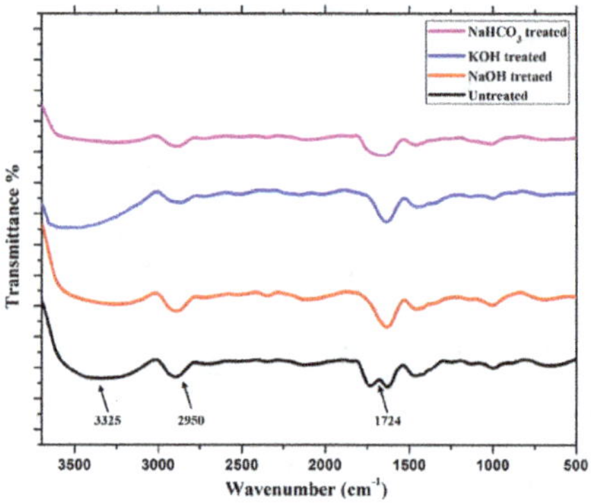

Fig 2. FTIR spectra of untreated and alkali-treated hemp fibre fabrics

Effect of Alkaline treatment on flexural properties

Fig. 3 illustrates the variation in flexural properties across different surface-treated NFRP composites. The flexural stress versus strain curves for untreated hemp, 5% NaOH, 5% KOH, and 5% $NaHCO_3$-treated hemp/epoxy composites is shown in Fig. 3(a). The flexural strength of the $NaHCO_3$ HE was found to be higher than that of the other composites. The flexural strength and modulus of composites fabricated using $NaHCO_3$-treated hemp fibres are higher by ~26% and ~32%, respectively, than untreated hemp/epoxy composites. This results from the removal of hemicellulose and surface impurities, such as wax, and might significantly increase surface roughness compared to other treatments. The microfibrils became more efficient at densely reorienting, leading to closer packing of the cellulose chains and improved fibre strength and flexural properties. According to Das et al. [6], alkaline and sodium bicarbonate treatment increased flexural strength by 18% and 33%, respectively, for 9% concentrated solutions. While NaOH treatment caused fibre degradation, $NaHCO_3$ treatment eliminated the hemicellulose and wax from the fibre.

Materials Research Proceedings 65 (2026) 30-34 https://doi.org/10.21741/9781644904138-5

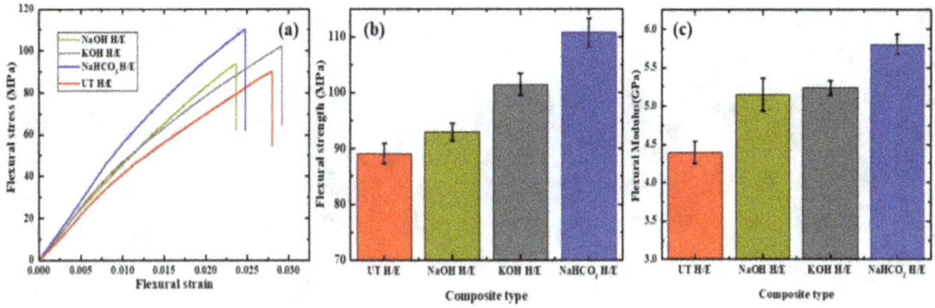

Fig 3. (a) Flexural stress-strain relationship (b) Flexural strength of different composites (c) Flexural modulus of different composites.

Effect of Alkaline treatment on ILSS properties

Fig. 4 illustrates the variation in interlaminar shear properties across different surface treated NFRP composites. The interlaminar shear strength of the 5% $NaHCO_3$ treated sample was found to be higher than that of the other hemp/epoxy composites. The interlaminar shear strength of composites fabricated using $NaHCO_3$-treated hemp fibres is significantly higher by ~29%, than untreated hemp/epoxy composites. The ILSS of $NaHCO_3$ HE composites was higher than NaOH HE and KOH HE composites by ~14% and ~8%, respectively. The sodium bicarbonate treatment has been removed the cementing layer, hence improving the ILSS of hemp-epoxy composites. This improvement can be attributed to an increase in the amount of contact surface between the fibre and matrix, which facilitated mechanical interlocking. FTIR analysis revealed that the sodium bicarbonate treatment eliminated non-cellulosic elements from fibres.

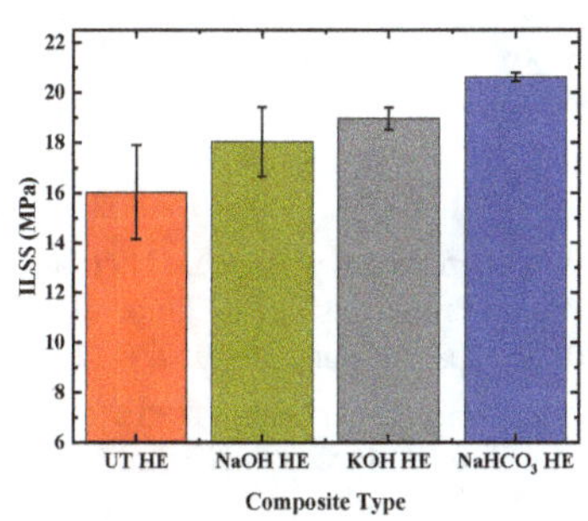

Fig 4. ILSS of untreated and different alkali treated hemp fibre/ epoxy composites

Fractography

This observation clearly suggests that $NaHCO_3$ treatment effectively removes the hemicellulose and lignin coatings from the cellulose surfaces of hemp fibres. The rough surface of hemp fibre after treatment with $NaHCO_3$ is shown in Fig 5 (a). The surface treatment with NaOH and KOH helps to partially remove the hemicellulose and lignin contentThe 5% $NaHCO_3$ NFRE composite, as shown in Fig. 5(b), exhibits strong fibre/matrix interfacial adhesion. . The river lines shown in Fig 5(c), which are a commonly observed features in epoxy-based composites, were found in almost all treated and untreated composites. This indicates the brittle failure of hemp/epoxy composites.

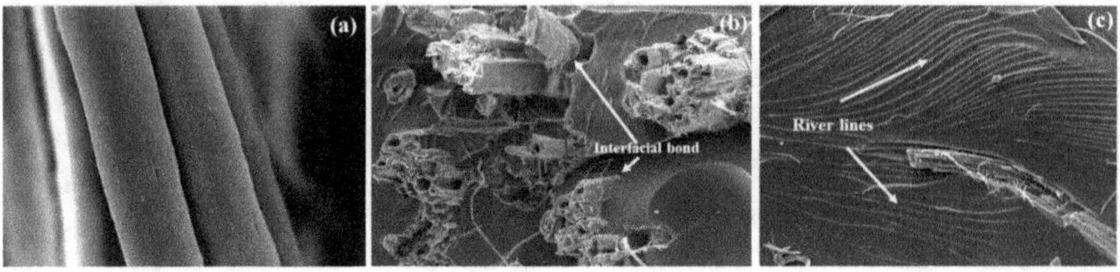

Fig 5. SEM micrograph of (a) $NaHCO_3$ treated hemp fibre surface and (b) a region with good interfacial bonding in $NaHCO_3$ HE composite, and (c) fractured surface of $NaHCO_3$ HE composite depicting river lines

Conclusion

Hemp fibres' surface treatment with NaOH, KOH, and NaHCO3 removed the hemicellulose and lignin contents from the hemp fibres surfaces and improved mechanical performance of hemp fibre/epoxy composites. Among the considered alkaline treatments, the hemp/epoxy composites fabricated using the hemp fibres treated with 5wt% NaHCO3 exhibited superior flexural and ILSS. The NaHCO3 HE composites demonstrate a notable ~26% improvement in flexural strength and a ~32% increase in flexural modulus. Furthermore, the ILSS, a critical metric for assessing fibre-matrix adhesion, enhanced by ~ 29%. These increments are primarily attributed to the effective removal of surface impurities and the partial fibrillation of the natural fibres, which increases the effective surface area and facilitates superior mechanical interlocking within the polymer matrix. To further advance the mechanical performance of hemp/epoxy composites beyond alkaline pre-treatments, nanofillers modification or a polymer sizing can be explored.

References

[1] P. Mutjé, J. Gironès, A. Lòpez, M. F. Llop, and F. Vilaseca, "Hemp Strands: PP Composites by Injection Molding: Effect of Low Cost Physico-chemical Treatments," *J. Reinf. Plast. Compos.*, vol. 25, no. 3, pp. 313–327, 2006. https://doi.org/10.1177/0731684406059784

[2] A. Shahzad, "Hemp fibre and its composites – a review," *J. Compos. Mater.*, vol. 46, no. 8, pp. 973–986, 2012. https://doi.org/10.1177/0021998311413623

[3] S. O. Amiandamhen, M. Meincken, and L. Tyhoda, "Natural Fibre Modification and Its Influence on Fibre-matrix Interfacial Properties in Biocomposite Materials," *Fibres Polym.*, vol. 21, no. 4, pp. 677–689, Apr. 2020. https://doi.org/10.1007/s12221-020-9362-5

[4] M. A. Sawpan, K. L. Pickering, and A. Fernyhough, "Effect of fibre treatments on interfacial shear strength of hemp fibre reinforced polylactide and unsaturated polyester composites," *Compos. Part Appl. Sci. Manuf.*, vol. 42, no. 9, pp. 1189–1196, 2011. https://doi.org/10.1016/j.compositesa.2011.05.003

[5] F. Sarker, M. Z. Uddin, K. Sowrov, M. S. Islam, and A. Miah, "Improvement of mechanical and interfacial properties of hot water and sodium bicarbonate treated jute fibres for manufacturing high performance natural composites," *Polym. Compos.*, vol. 43, no. 3, pp. 1330–1342, 2022. https://doi.org/10.1002/pc.26451

[6] P. P. Das *et al.*, "Environmentally sustainable chemical treatment of plant fibres for improved performance of polymeric composites," *Polym. Compos.*, vol. 43, no. 10, pp. 7155–7169, 2022. https://doi.org/10.1002/pc.26779

Materials Research Proceedings 65 (2026) 35-40 https://doi.org/10.21741/9781644904138-6

From forest waste to functional adsorbent: NaOH-activated pine needle biochar for Rhodamine B remediation

Bhavana V P[1,a], Neelaambhigai MAYILSWAMY[1,b], Balasubramanian KANDASUBRAMANIAN[1,c*], Rutuja Vasant SALUNKE[2,d]

[1]Department of Metallurgical and Material Engineering, Defence Institute of Advanced Technology (DU), Ministry of Defence, Girinagar, Pune, Maharashtra 411025, India

[2]Department of Microbiology, STES Sinhgad College of Science, Ambegaon, Pune, Maharashtra 411041, India

[a]bhavanavp23@gmail.com, [b]neelaambhigai.mayilswamy@gmail.com, [c]meetkbs@gmail.com, [d]salunkherutuja12@gmail.com

Keywords: Pine Needle, NaOH-Activated Pine Needle Biochar, Rhodamine B Dye, Adsorption Kinetics, Adsorption Isotherms, Thermodynamics, Regeneration Studies

Abstract. This study explores NaOH-activated pine needle biochar (NaOH-PNB) for RhB removal from wastewater. Pine needles were pyrolyzed at 500°C, treated with 2 M NaOH, and re-pyrolyzed at 800°C to improve surface properties. The material showed a mesoporous structure suitable for adsorption. Maximum removal was achieved in 5 h, with PSO kinetics and a capacity of 4.08mg.g^{-1}. The process was spontaneous and endothermic, driven by electrostatic interactions and hydrogen bonding. The adsorbent was reusable for up to 3 cycles using ethanol. NaOH-PNB showed better performance than other bio-adsorbents, supporting sustainable wastewater treatment.

1. Introduction

Water pollution management is a significant global issue, primarily due to human activities contaminating aquatic environments, which threatens public health and water quality. Addressing this requires the development of effective and economical materials. Synthetic dyes like Rhodamine B (RhB), used in industries such as textiles, are major pollutants in wastewater, posing health risks and environmental harm. Various treatment methods exist, but processes of adsorption are gaining popularity due to their economic viability and efficiency in removing dyes without by-products. While activated carbon is a common adsorbent, its high-cost limits usage. Biochar, produced from biomass pyrolysis, has emerged as a sustainable and efficient alternative for wastewater treatment, capable of adsorbing a variety of contaminants due to its porous structure and functional groups[1]. Pine trees, with around 130 species, are found globally, particularly in the Himalayan ecosystem of Uttarakhand, India, which features over 340,000 hectares of pine forests producing around 2.06 million tonnes of pine needles annually. While highly flammable and contributing to environmental degradation, pine needles can be converted into biosorbents through thermochemical methods, promoting a sustainable bioeconomy. Specifically, NaOH-activated pine needles can efficiently remove the harmful dye Rhodamine B (RhB) from wastewater. This study aims to synthesize and assess NaOH-activated pine needle biochar (NaOH-PNB) for RhB remediation, evaluating factors like contact time(kinetics), initial dye concentration(isothermal), temperature(thermodynamics), adsorbent dosage, and pH, while also exploring the adsorption mechanism and options for reuse.

2. Methodology and Materials Used

2.1 Materials and Reagents Employed: Pine needles were sourced from Raigad district, Maharashtra, India. The biomass was subjected to drying at 100 °C for 24 h, followed by grinding

Materials Research Proceedings 65 (2026) 35-40 https://doi.org/10.21741/9781644904138-6

to obtain a fine powder, and stored in airtight containers. Rhodamine B was obtained from Sisco Research Laboratories, NaOH from Sigma Aldrich in Bangalore, glacial acetic acid and hydrochloric acid from Loba Chemie, and pure ethanol from Zenith Bio-Chemical Industries.

2.2 Fabrication of NaOH-Activated Biochar from Pine Needles: To produce Pine Needle Biochar (PNB), 10 grams of dried biomass were pyrolyzed in a stainless-steel furnace under inert conditions. The biochar was ground, mixed with 2M NaOH for 24 h, then filtered and dried at 105°C. A second pyrolysis occurred in nitrogen at 800°C for 2 hours. Multi-step pyrolysis separates carbonization and activation, specific surface area is enhanced compared to single-step pyrolysis. This expanded internal structure yields superior physicochemical properties and substantially higher adsorption capacity, making it highly effective for dye removal. The final product, NaOH-PNB, was rinsed and dried for 24 h(Fig.1).

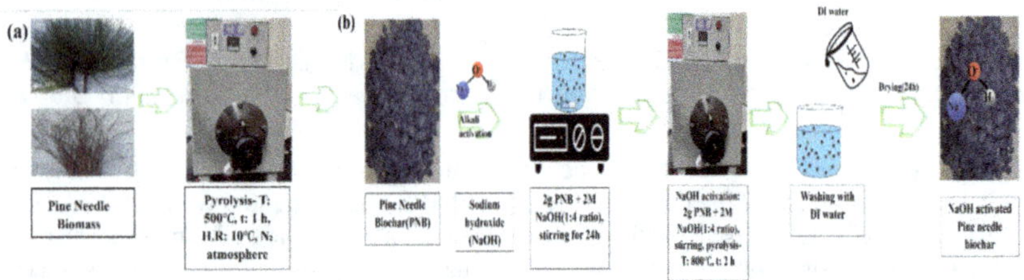

Figure 1: Process flow schematic depicting the synthesis procedures for (a) PNB and (b) NaOH-PNB adsorbents

NaOH-PNB was characterized using FTIR, Raman, XRD, SEM-EDX, BET-BJH, and XPS to analyze structure, morphology, surface area, and chemical states. For adsorption, 20 mg of adsorbent was added to 20 mL of 15 ppm RhB (pH 4.76) and studied up to 420 min, with concentration measured at 554 nm. Effects of dosage, time, pH, temperature, and concentration were evaluated in triplicate. Adsorption capacity and removal efficiency were calculated using standard equations.[2].

$$q_e = \frac{(C_0 - C_e).V}{m} \quad (1) \qquad R = \frac{(C_0 - C_e) \times 100}{C_0} \quad (2)$$

Where, C_o= The dye's initial concentration in[$mg.L^{-1}$], q_e= the amount of RhB taken up by the NaOH-PNB adsorbent at equilibrium in [$mg.g^{-1}$], C_e= Concentration of RhB at the equilibrium stage in[$mg.L^{-1}$], V = Volume of dye solution(RhB) in [L], m= biochar weight in [g], R = removal efficiency of dye in %.

2.3 Kinetics, isotherms, and thermodynamics of RhB adsorption: RhB adsorption kinetics were analyzed using PFO, PSO, IPD, and Elovich models. Isotherms were studied using Langmuir, Freundlich, Temkin, Halsey, D-R, and Sips models to evaluate capacity and interactions. The D-R model showed $E < 8$ $kJ.mol^{-1}$, indicating physisorption.

2.4 Regeneration Studies: Three adsorption-desorption cycles were performed to assess reusability. Regeneration was done using ethanol for 6 h, and RhB removal was measured by UV-Vis analysis of the supernatant.

3. Analysis of Experimental Findings

3.1 Characterization of NaOH-PNB: FTIR spectra of NaOH-PNB before adsorption showed peaks at 2978(C-H), 1695(C=O), 1450(C=C), 1186(C-C), 1077(C-O), 958(C=O), 870(aromatic C-H), and 473 and 414[cm^{-1}](Si-O-Si). Peak shifts from 2978 to 3019, 958 to 981, 870 to 840, and 473 to 444[cm^{-1}] indicate structural modifications after adsorption. Retention of peaks at 1695and ~444 confirms the presence of C=O and Si-O-Si groups post adsorption. These shifts and new peaks confirm RhB uptake onto NaOH-PNB, mainly governed by π-π interactions between dye molecules and biochar aromatic structure. After RhB adsorption, new peaks appeared at 3450(N-H), 1563(C=N), 653(C-H), along with 981, 840[cm^{-1}](C=O, C-H)[3] These shifts and new peaks confirm RhB uptake onto NaOH-PNB, mainly governed by π-π interactions, n-π interactions and electrostatic interactions. π-π interactions occur between the aromatic π-electron system of the RhB dye and the π electrons present in the aromatic domains of the biochar structure. Fig. 2b shows XRD peaks at 22.8° and 35.7° for cellulose I, 29.2° for carbon, 20-28° for amorphous graphite, and 43.1° for turbostratic carbon. Peak intensity decreased after adsorption, indicating reduced crystallinity. Raman(Fig. 2c) shows D(1353 cm^{-1}) and G(1588 cm^{-1}) bands, with I_D/I_G=0.852 and L_a=22.59[nm], indicating defects and partial graphitization. SEM (Fig. 2d) shows increased roughness and pores after activation, with a cracked surface after RhB adsorption, confirming dye uptake[4]. EDX(Fig.2e,f) shows C(69.02%), O(23.72%), and Na(5.52%), with C increasing to 82.51% and O decreasing to 13.55% after adsorption. N_2 isotherm (Fig. 2g,h) confirms mesoporosity, with BET area 5.923 [m^2/g] and pore size 3.841[nm]. XPS (Fig. 2i,j) shows C 1s(69.34%), O 1s(13.05%), Cl 2p(8.6%), Na 1s(8.45%), and Mg 1s(0.57%), confirming oxygen functional groups.

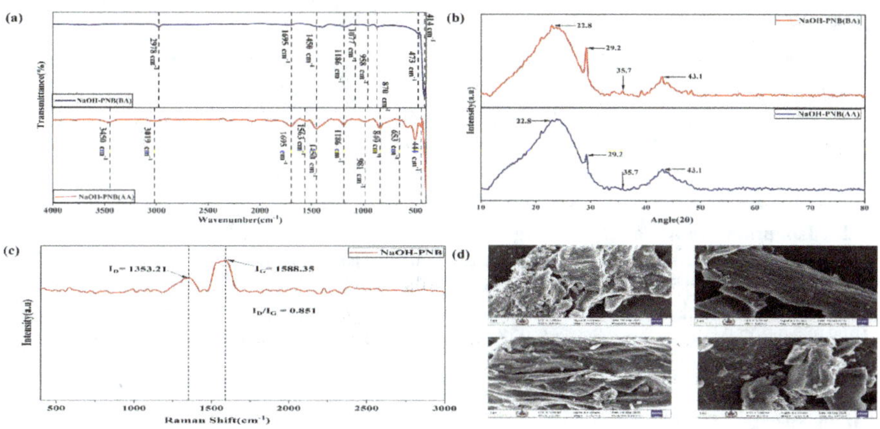

Materials Research Proceedings 65 (2026) 35-40
https://doi.org/10.21741/9781644904138-6

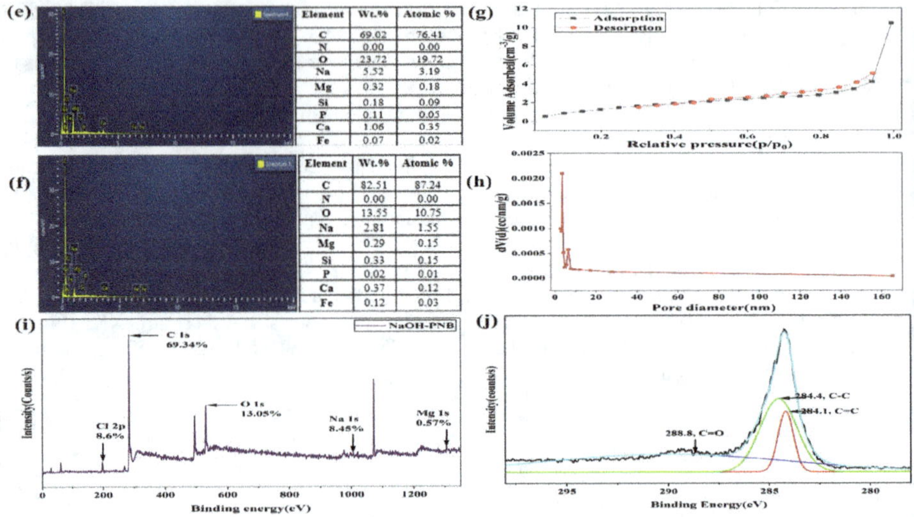

Fig 2: (a)FTIR, (b)XRD, (c)Raman (before RhB adsorption), (d)SEM, (e,f)EDX, (g,h)BET-BJH, and (I,j)XPS analyses of NaOH-activated pine needle biochar (NaOH-PNB) before (BA) and after (AA) RhB adsorption

3.2 Adsorption Studies of RhB Using NaOH-Derived Pine Needle Biochar

3.2.1 Influence of Contact Time and Kinetic Analysis: RhB adsorption on NaOH-PNB (Fig. 3a) was rapid initially (15-300 min) and reached equilibrium at 300 min. The PSO model(R^2=0.9956) best fit the data, with q_e=6.25[mg.g^{-1}](exp: 6.28[mg.g^{-1}]), indicating chemisorption. IPD(R^2=0.8924) and Elovich(R^2=0.9771) showed secondary effects, while PFO(R^2=0.9049) confirmed poorer fitting.

3.2.2 Influence of Initial RhB Concentration and Isotherm Studies: RhB adsorption (0.14-15[mg.L^{-1}]) increased capacity to 3.65[mg.g^{-1}], while removal efficiency decreased from ~80% to 24.33%. Langmuir fit best (R^2=0.9947), with q_m=4.8309[mg.g^{-1}] and K_L=0.67 L[mg^{-1}], indicating monolayer adsorption[5]. Temkin(R^2=0.936, E = 3837.1[J.mol^{-1}]) and D-R(R^2=0.978, E=4.0825[kJ.mol^{-1}]) confirmed physisorption. Freundlich(R^2=0.986), Halsey (R^2=0.986), and Sips(R^2=0.991) showed favorable adsorption with R_L=0-1.

3.2.3 Influence of Adsorbent Dosage: Increasing NaOH-PNB dosage (0.5-5[g/L], Fig. 3e) raised removal efficiency from 35.25% to 66.48%. However, adsorption capacity decreased at higher doses due to particle aggregation and reduced effective surface area.

3.2.4 Influence of Temperature and Thermodynamics: Increasing temperature (25-45°C, Fig. 3f) raised removal from 51.94% to 63.70% and capacity from 7.79 to 9.55[mg/g], indicating an endothermic process. Thermodynamics(Fig. 3g) showed negative $\Delta G°$(spontaneous) and positive $\Delta H°$ and $\Delta S°$, indicating endothermic adsorption with increased disorder.

3.2.5 Influence of Solution pH: RhB adsorption was highest at low pH and decreased from 36.28% to 27.04% over pH 2-12(Fig. 3h). Higher adsorption at acidic pH is due to dominant chemisorption, while at higher pH mixed interactions occur. Above pKa 3.7, RhB forms zwitterions and dimers, limiting pore access[6].

3.2.6 Adsorption Mechanism: RhB uptake occurs via π–π, electrostatic, and n–π interactions, along with pore filling. Its size is smaller than the pore diameter (3.841 nm), allowing easy diffusion and trapping in NaOH-PNB (Fig. 4a).

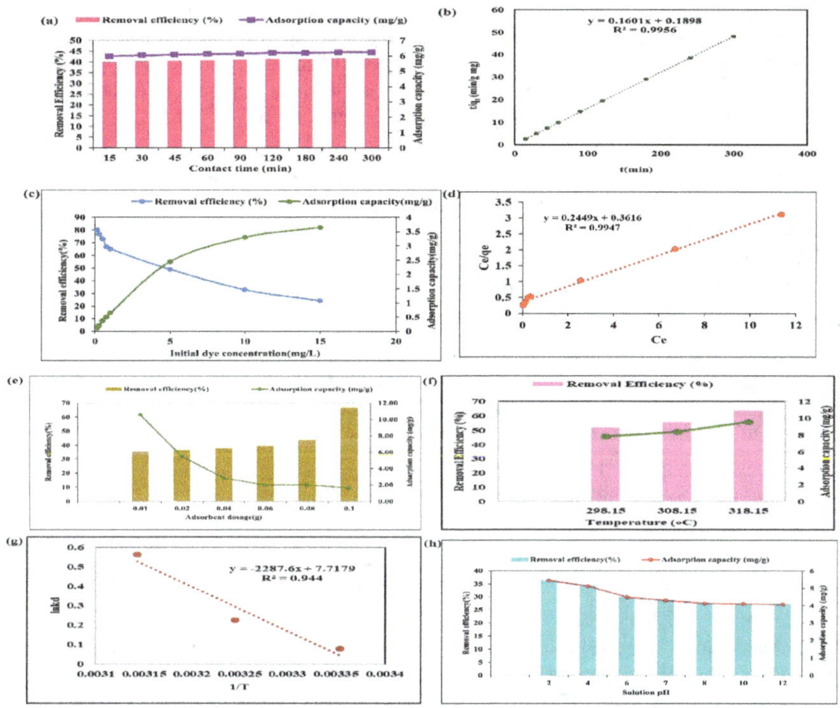

Fig.3: (a)Effects of contact time, (b)adsorption kinetics (PSO), (c)initial RhB concentration, (d)adsorption isotherms (Langmuir), (e)adsorbent dosage, (f)temperature, (g)thermodynamics, and (h)solution pH on RhB adsorption by NaOH-activated pine needle biochar (NaOH-PNB).

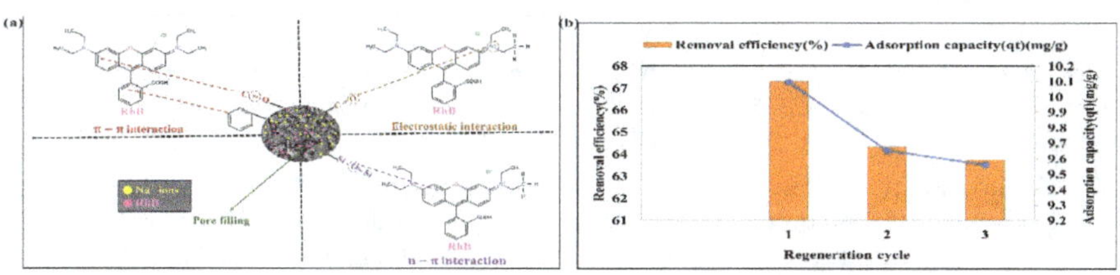

Fig.4: (a)RhB adsorption mechanism on NaOH-PNB and (b)adsorbent regeneration performance.

3.3 Desorption and Regeneration Studies

Regeneration studies using 0.1 N ethanol over three adsorption-desorption cycles (Fig.4b) showed a slight decrease in removal efficiency from 67.31% to 63.72%. This reduction is attributed to partial structural changes and loss of active sites; however, NaOH-PNB retained good reusability, confirming its practical applicability[7].

Conclusion

NaOH-PNB from pine needles was developed as an eco-friendly adsorbent for RhB removal. Two-stage activation improved structure and performance, with optimal removal at pH 2-4.76 and up to 45°C. Adsorption followed PSO kinetics and Langmuir isotherm, with a capacity of 4.08[mg/g] and endothermic behavior. The material was reusable and supports sustainable wastewater treatment, though scale-up and real wastewater studies are needed.

Acknowledgements

The author would like to thank Dr. B.H.V.S Narayana Murthy, Vice-Chancellor of DIAT (DU), Pune, for his constant support and motivation. The first author would like to thank Ms. Bhavana Shanmughan for her constant support throughout the course of this technical work. The first author is also thankful to Ms. Shruti Gupta, Ms. Payal Varma, Ms. Manisha Singh, Ms. Kshitija Vaidya, and Ms. Deepali Kadam for their support throughout this work.

Reference

[1] N. Mayilswamy, B. Kandasubramanian, T. Warjurkar, S. Chame, and S. Shirin, "Optimization and prediction of Rhodamine B uptake onto alkali-functionalized sugarcane bagasse biochar: Krill Herd algorithm-based ANN modelling approach," *Environ. Sci. Pollut. Res.*, Jul. 2025. https://doi.org/10.1007/s11356-025-36680-1

[2] A. Venkatesan *et al.*, "High adsorption capacities of rhodamine B dye by activated carbon synthesized from cotton stalks agricultural waste by chemical activation," *Ceram. Int.*, vol. 51, no. 10, pp. 13345–13354, Apr. 2025. https://doi.org/10.1016/j.ceramint.2025.01.178

[3] T. A. Saleh and I. Ali, "Synthesis of polyamide grafted carbon microspheres for removal of rhodamine B dye and heavy metals," *J. Environ. Chem. Eng.*, vol. 6, no. 4, pp. 5361–5368, Aug. 2018. https://doi.org/10.1016/j.jece.2018.08.033

[4] B. Acemioğlu, "Removal of a reactive dye using NaOH-activated biochar prepared from peanut shell by pyrolysis process," *Int. J. Coal Prep. Util.*, vol. 42, no. 3, pp. 671–693, Mar. 2022. https://doi.org/10.1080/19392699.2019.1644326

[5] D. Paluch, A. Bazan-Wozniak, R. Wolski, A. Nosal-Wiercińska, and R. Pietrzak, "Removal of Methyl Red from Aqueous Solution Using Biochar Derived from Fennel Seeds," *Molecules*, vol. 28, no. 23, p. 7786, Nov. 2023. https://doi.org/10.3390/molecules28237786

[6] A. O. Adeola, M. Cui, and R. Naccache, "Rhodamine B sequestration using acid-precipitated and microwave-treated softwood lignin: Comparative isotherm, kinetics and thermodynamic studies," *Environ. Technol. Innov.*, vol. 32, p. 103419, Nov. 2023. https://doi.org/10.1016/j.eti.2023.103419

[7] N. Mayilswamy, B. Kandasubramanian, M. Nannaware, and P. M. Gore, "ANN Modeling for Rhodamine B Adsorption Using Pristine and NaOH-Activated Mesoporous Sewage Sludge Biochars: Kinetic, Isotherm, Thermodynamic, and Regeneration Studies," *Ind. Eng. Chem. Res.*, vol. 64, no. 6, pp. 3288–3316, Feb. 2025. https://doi.org/10.1021/acs.iecr.4c02105

Cyclic deformation behavior and microstructural evolution in nickel nanowire with internal void

Natraj YEDLA*, Kumud Kant MEHTA

Department of Metallurgical and Materials Engineering, NIT Rourkela-769008, India

yedlan@nitrkl.ac.in

Keywords: Molecular Dynamics, Strain-Control, Cyclic Deformation, Defects, Dislocations, Strain Localization

Abstract. In this study, the response to strain-controlled (SC) cyclic deformation behavior of nickel nanowire (NW) containing a void is investigated. The void is located at the center of the NW internally (coordinates of the center of the void (x, y, z) = 500 Å, 50Å, 50Å, radius = 15 Å). At first, the NW is subjected to uniaxial tensile deformation at a constant strain rate of 10^9 s^{-1} and at a temperature of 300 K to determine the mechanical properties. Peak stress = 13.1 GPa, yield strain = 0.162, and fracture strain = 0.193 are observed in the NW. Based on the tensile test results, the NW is then subjected to compression-tension cycles in the elastic strain amplitude range of 0.06–0.12 at a temperature of 300 K. The microstructural analysis during the tensile deformation shows that plastic deformation is localized near the void. The void acts as a source of dislocations. The expansion of the void takes place by the continuous emission of dislocations with the progress of deformation. Collapse of the void occurs at a strain of e = 0.186, followed by the failure of the NW. The SC cyclic behavior stress-strain cycles show an increase in the width of the loops with increasing cycles due to cyclic softening. Fracture of the NW is not observed even after 1000 cycles of cyclic deformation. The microstructural features during SC cyclic deformation reveal the formation of dislocation loops at the void surface and deformation bands on the NW surface.

1. Introduction

Nanowires (NWs) have superior optical, electrical, magnetic, and mechanical properties, which make them suitable for different nanoscale devices [1–3]. The mechanical behavior and underlying mechanism are vital for device applications [4]. Although several in-situ experiments have revealed the strength and fracture behavior of NWs [5–7], the tests suffer from drawbacks [8]. So, alternate methods like molecular dynamics (MD) simulations have been used to gain insight into the nanoscale properties [8,9]. There is a vast amount of literature reported using MD simulations on the mechanical response of metallic NWs by considering different NW sizes, strain rates, defects, and temperatures [10,11].

Understanding the response to cyclic loading is also important in view of the applications [12–14]. The in-situ fatigue tests report that failure of nickel NWs occurs after 375 cycles [12]. The MD fatigue studies on pre-cracked Al nano-chips report failure after 13 cycles [13]. The microstructural analysis reveals the formation of nano-voids, dislocation cells, and persistent slip bands on the nano-chips [13]. Investigations on the failure due to fatigue in nanoscale nickel single crystal reveal that fatigue crack growth rate is not dependent on the crystallographic orientation [14]. Chen et al. [15] report that nickel base single crystal superalloy exhibits excellent fatigue mechanical properties.

It can be seen from the above literature, and to the author's knowledge, very few fatigue failure studies are reported at the nanoscale [15,16]. Defects such as cracks and voids, which act as stress concentrators, lower the material's mechanical performance and can form during manufacturing [17]. On the other hand, the effect of internal defects on fatigue behavior is seldom reported in nickel NWs. Moreover, MD has proven to be a vital computational tool for investigating the properties at the nanoscale [11]. So, in the present study, the author investigates the role of a void on the fatigue behavior and microstructural evolution during strain-controlled (SC) cyclic deformation in nickel NW.

2. Simulation details

2.1 Methodology and interatomic potentials

The MD simulations of uniaxial tensile and SC cyclic deformation tests of defect nickel NW are carried out on the LAMMPS platform [18]. The forces between the atoms are estimated using the embedded atom method (EAM) potential [19]. Researchers have used the potential for plastic deformation studies and estimation of physical properties [20].

2.2 Models and deformation parameters

The model nickel NW used for the present study has the dimensions of 100 Å (x-axis [100]) × 1000 Å (y-axis [010]) × 100 Å (z-axis [001]), respectively. Internal single void (diameter: 30 Å; the position of the void (500 Å (y-axis); 50 Å (x-axis); 50 Å (z-axis)) is introduced in the NW by deleting atoms (Fig. 1a). After the initial construction of the NW relaxation is done using conjugate gradient method [21,22]. Thereafter, equilibration is performed at 300 K temperature using the Nose-Hoover thermostat [23]. The NW is subjected to a tensile test along the y-direction [0 1 0] (periodic boundary (y-axis) and non-periodic (x- [1 0 0] and z- [0 0 1]axes)) at a constant strain rate (10^9 s^{-1}) and temperature (300 K). Thereafter, cyclic strain is applied at constant strain amplitudes of 0.06, 0.08, and 0.12 along the y-axis (periodic boundary (y-axis) and non-periodic (x- and z-axes)). NVT integrator (N = number of atoms, V = Volume, and T = temperature are constant) and integration timestep of 1fs is used in all the above simulations. The temperature is maintained at 300 K during cyclic tests using a Nose-Hoover thermostat [23]. The compression-tension cycle corresponding to a strain amplitude of 0.06 is shown in Fig. 1b.

2.3 Visualization tool and analysis

Simulation tool OVITO [24] is used for identifying the deformation mechanisms and lattice order/disorder in the nickel NW during tensile and cyclic deformations. The dislocation extraction algorithm (DXA) modifier [25] is used to identify dislocations generated during plastic deformation. The strain localization is studied using the atomic shear strain analysis tool, which uses von Mises shear strain expression [26]. Virial stress [27] is used for stress calculations and engineering strain for the estimation of strain [22]. Several MD simulation studies on nanocrystalline materials report the use of the above tools for identifying crystal defects [28], deformation mechanisms [29], and strain localization [26] during deformation.

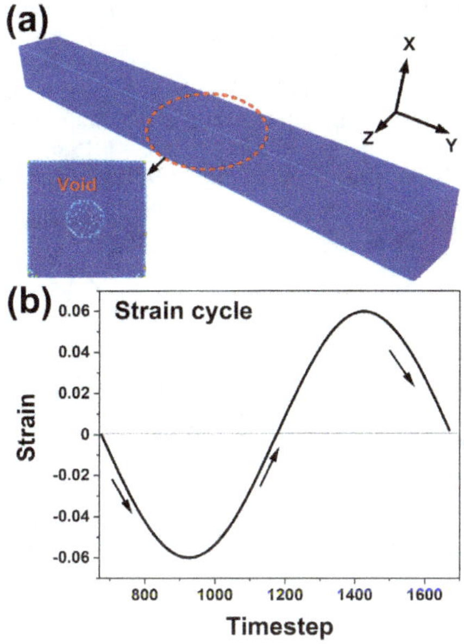

Fig. 1: *(A) Atomic snapshots of nickel NWs having void (sectional view (X-Z plane)); (b) Typical compression-tension strain cycle (arrows indicate the direction of strain) used in the present investigation.*

3. Results and Discussion

3.1 Stress-strain curves: tensile behavior

Fig. 2 shows the stress-strain response of nickel NW with the void. The microstructural changes occurring in the NW with a void are captured at specific strains. The NWs show linear elastic behavior up to a maximum stress, and thereafter due to yielding, a sharp drop in the stress is observed. Similar stress-strain behavior is reported in the studies on metallic NWs [30]. The peak stress values are found to be 13.1 GPa (e = 0.162) in the NW. The values in the parentheses indicate yield strains. The yield strength in the range of 8 GPa–12 GPa has been reported for perfect nickel NWs (strain rate range: 10^8 s^{-1}–10^{12} s^{-1}; temperature: 300 K) [21]. The present study results are different compared to the literature, which could be due to the size effect and also the interatomic potential. DXA analysis has been carried out in NW with a void to investigate the deformation mechanisms. The DXA analysis at strain e = 0 shows no sign of dislocations in the NW, indicating it is pristine. The void surface gets slightly distorted (e = 0.16). Shockley partial dislocation loop (zoomed region; Burgers vector: 1/6 [2-11]) emanates from the void surface (e = 0.162), which results in a drop in stress. The void surface is the source of dislocation due to high stress concentration [16]. Thereafter, more Shockley partials emanate from the void surface, resulting in void growth and volume fraction with increasing strain (e = 0.166–e = 0.168). The stress continually decreases due to a burst of dislocations [31]. Ahn et al. [32] demonstrated that dislocation loops emit from the void surface and are energetically favorable in face-centered cubic metals. Pohjonen et al. [33] and Traiviratana et al. [34] also report nucleation of dislocations from the void surface and void growth using MD simulation studies in Cu. Furthermore, Bobyley et al.

[35] have shown that the initiation of dislocation semi-loops is energetically favorable and depends on the grain size and magnitude of stress in nanocrystalline materials. In the present study, apart from Shockley partials, few perfect dislocations (Burgers vector: 1/2<110>; blue color) and strain rod (Burgers vector: 1/6<110>; pink color) dislocations can also be seen (e = 0.172; e = 0.186). The NW defect mesh (Fig. 1c) snapshot captured at the strain of e = 0.174 shows localized deformation at the void site, and final fracture at strain e = 0.193. Liu et al. [36] also report neck formation before fracture in the MD studies on the superplastic behavior of Au NWs.

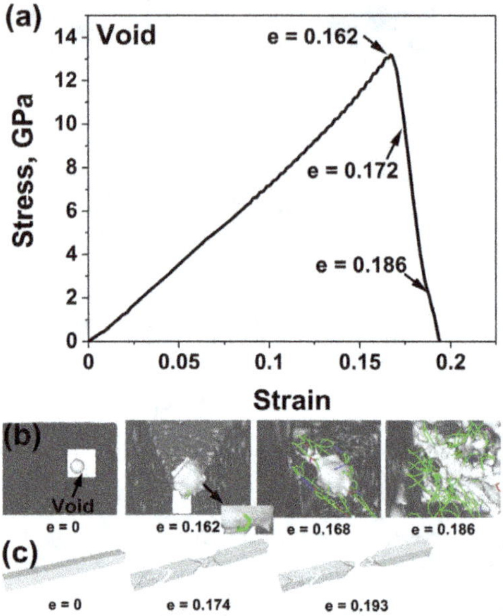

Fig. 2: *Stress-strain response of nickel NWcontaining void (strain rate of 10^9 s^{-1}; temperature: 300 K). The DXA analysis and defect mesh snapshots of the void NW are shown at different strains till fracture. Shockley partial (green color), perfect dislocation (blue color), and stair rod dislocation (pink color).*

3.2 Cyclic stress-strain curves (compression-tension cycle) and microstructural evolution
Fig. 3 shows the cyclic stress-strain curves (hysteresis loops) obtained at the 1st, 100th, and 1000th cycles (strain amplitudes: 0.06, 0.08, and 0.12) in the NW with void. The strains used in the cyclic deformation correspond to elastic deformation (Fig. 2). The cyclic stress-strain curves show asymmetric behavior at all the strain amplitudes. The compressive peak stress (~7GPa) is greater than the tensile peak stress (~6 GPa) for all the cycles during compression followed by the tensile cycle. Wang et al. [37] also report asymmetric stress-strain hysteresis loops in single crystal nickel superalloy subjected to low cycle fatigue which is correlated to the dislocation microstructures. Further, Jabbari et al. [38] report that the asymmetric stress-strain behavior changed to symmetric in the studies of AZ31B magnesium alloy as the temperature increased from 100 °C to 200 °C [38].

The width of the hysteresis loops increases as the strain cycles increase due to the cyclic softening. The microstructural analysis can explain the reason behind the softening process, which will be discussed in the following paragraph. Akbarian and Dehgani [13] also report cyclic softening behavior in pre-cracked aluminum nano-chips during strain cycling (tension-compression cycles). Further, they report an increase in fatigue life during the tension-compression cycle (13 cycles) than the tension-tension cycle (4 cycles) due to the de-voiding process and formation of the dislocation substructure. In the present study, fracture of the NWs is not observed because the strain amplitudes are low. The difference in the compressive and tensile peak stress values for the different defect NWs is slight. This is also the same with increasing the amplitude and cycles. A similar behavior has been reported in the fatigue studies of defect aluminum nano-chips [13].

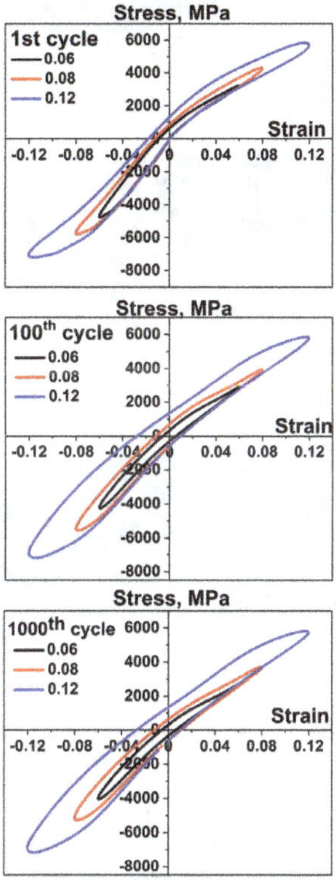

Fig. 3: Plot of hysteresis loops at different strain amplitudes and cycles: (a) 1st cycle, (b) 100th cycle, and (c) 1000th cycle.

The microstructural changes in NW subjected to a strain amplitude of 0.06 are shown in Fig. 4. After the 1st cycle (Fig. 4a) there is no significant surface deformation, which is evident from the color contour. However, at the void site, the emission of Shockley partial dislocation loop can be seen. The Burgers vector of the dislocation is identified as 1/6 [-2-1-1]. At the 100th cycle (Fig. 4b), the NW surface seems to show deformed regions and also the void is no more spherical. With further increasing the cycles to 1000 (Fig. 4c), more deformed regions (deformation bands) are observed on the NW surface, and also multiple dislocation loops are emitted from the void surface contributing to the plastic strain of the NW. The bands are observed to be parallel to the slip lines and hence are of DBI type [39]. Burque et al. [40] also report the formation of deformation bands on the surface of nickel single crystal ($[\bar{1}\bar{1}1]$ orientation). The void gets distorted due to dislocation emission from the surface. The authors report that similar deformation features are observed even at higher strain amplitudes (not shown in the present study).

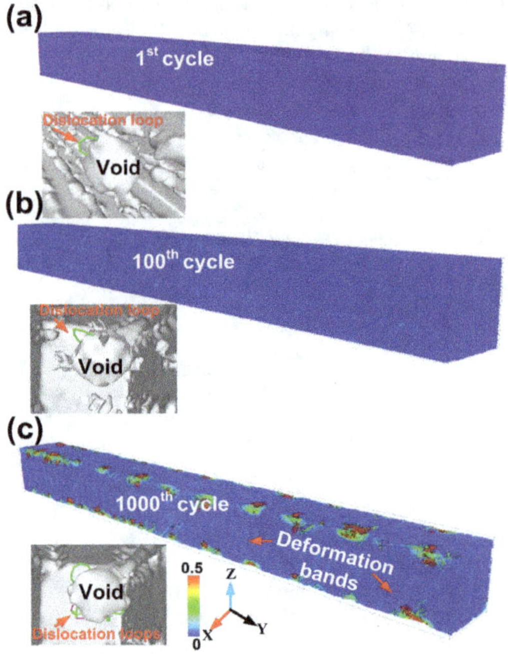

Fig. 4: Microstructural evolution at strain amplitude of 0.06 and at different cycles: a) 1st ; b) 100th; and c) 1000th.

3.3 Peak stress and plastic strain variation

Fig. 5a shows the variation of peak compressive (C) and tensile (T) stresses at different cycles and strain amplitudes. In Fig. 5a (strain amplitude: 0.06), the stress gradually decreases till the 1000th cycle in the compression cycle, whereas in the tensile cycle, there is a slight increase in the stress till the 10th cycle. Thereafter, a decrease in stress is observed till the 100th cycle. Upon further increase in the cycles, the stress remains almost constant till the 1000th cycle. The decrease in stress can be attributed to the plastic deformation by slip, causing softening of the NW. At strain amplitudes of 0.08 and 0.12, similar softening behavior is observed in the NW. It has been reported that strain-hardening behavior is observed in aluminum alloy [41]. In the cyclic deformation studies on metallic glasses, hardening is observed at low loads while softening at high loads [42]. Furthermore, Sha et al. [43] report that shear bands cause softening in Cu-Zr metallic glasses. Fig. 5b shows the plastic strain variation with increasing cycles at different strain amplitudes. The plastic strain (ε_p) is determined from the hysteresis loop, and the calculations have been done using a methodology similar to that adopted in the studies of Jian et al. [42]. In the NW, the plastic strain increases with the number of cycles at all the strain amplitudes. However, a plateau is observed after 100 cycles (strain amplitude: 0.12) and ten cycles (strain amplitude: 0.06 and 0.18), which could be due to a decrease in the glissile dislocations. The softening behavior is attributed to the plastic deformation by slip that has resulted in an increase in the plastic strain. The sources for dislocation nucleation is the void site as well as the NW surface edges. The cyclic softening behavior is also reported in nanoscale single crystal copper[44].

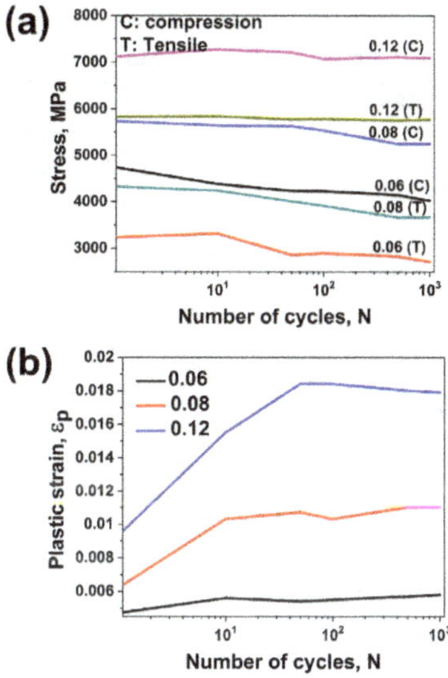

Fig. 5: (a) Peak stress variation with the number of cycles at different strain amplitudes; (b) Plastic strain variation with number of cycles at different strain amplitudes.

Conclusions

The following are the conclusions from the studies on the SC fatigue behavior and microstructural evolution in nickel NW containing a void.

a) The tensile test results show that NWs yield by dislocation slip caused by Shockley partial dislocation emission from the defect sites. The plastic deformation continues by the emission of dislocations from the surface edges. The plastic deformation is heavily localized at the defect site. Prior to fracture, necking is observed in the NW.
b) The SC cyclic tests reveal that the hysteresis loops show asymmetric stress-strain behavior. The hysteresis loops width increases, and the peak stress magnitudes decrease due to cyclic softening. Similar to that observed in tensile results, the defects act as a source of dislocations during cyclic deformation to initiate plastic deformation. Also, the formation of slip lines and slip bands is seen on the surface of the NWs.

Summary

The authors investigate the response to cyclic deformation behavior of nickel NW containing a void using MD simulations. The study reveals that the void acts as a source of dislocations and contributes to cyclic softening behavior. Under the present conditions, it is observed that failure of the NW is not observed even after 1000 cycles of cyclic deformation.

References

[1] Costa IM, Cunha TR, Cichetto Jr L, et al. Investigation on the optical and electrical properties of undoped and Sb-doped SnO2 nanowires obtained by the VLS method. Phys E Low-dimensional Syst Nanostructures. 2021;134:114856. https://doi.org/10.1016/j.physe.2021.114856

[2] Sarkar J, Khan GG, Basumallick A. Nanowires: properties, applications and synthesis via porous anodic aluminium oxide template. Bull Mater Sci. 2007;30(3):271-290. https://doi.org/10.1007/s12034-007-0047-0

[3] Stafford CA. Metal nanowires: quantum transport, cohesion, and stability. Phys status solidi. 2002;230(2):481-489. https://doi.org/10.1002/1521-3951(200204)230:2<481::AID-PSSB481>3.0.CO;2-M

[4] Tao W, Cao P, Park HS. Superplastic creep of metal nanowires from rate-dependent plasticity transition. ACS Nano. 2018;12(5):4984-4992. https://doi.org/10.1021/acsnano.8b02199

[5] Liu Z, Yuan X, Wang S, et al. Nanomechanical behavior of single taper-free GaAs nanowires unravelled by in-situ TEM mechanical testing and molecular dynamics simulation. Mater Sci Eng A. 2021;806:140866. https://doi.org/10.1016/j.msea.2021.140866

[6] Lu Y, Peng C, Ganesan Y, et al. Quantitative in situ TEM tensile testing of an individual nickel nanowire. Nanotechnology. 2011;22(35):355702. https://doi.org/10.1088/0957-4484/22/35/355702

[7] Zhu Y, Xu F, Qin Q, et al. Mechanical properties of vapor– liquid– solid synthesized silicon nanowires. Nano Lett. 2009;9(11):3934-3939. https://doi.org/10.1021/nl902132w

[8] McDowell MT, Leach AM, Gall K. Bending and tensile deformation of metallic nanowires. Model Simul Mater Sci Eng. 2008;16(4). https://doi.org/10.1088/0965-0393/16/4/045003

[9] Zhan HF, Gu Y, Yan C, et al. Numerical exploration of plastic deformation mechanisms of copper nanowires with surface defects. Comput Mater Sci. 2011;50(12):3425-3430. https://doi.org/10.1016/j.commatsci.2011.07.004

[10] Weinberger CR, Cai W. Orientation-dependent plasticity in metal nanowires under torsion: twist boundary formation and Eshelby twist. Nano Lett. 2010;10(1):139-142. https://doi.org/10.1021/nl903041m

[11] Huang D, Zhang Q, Qiao P. Molecular dynamics evaluation of strain rate and size effects on mechanical properties of FCC nickel nanowires. Comput Mater Sci. 2011;50(3):903-910. https://doi.org/10.1016/j.commatsci.2010.10.028

[12] Zhang H, Jiang C, Lu Y. Low-cycle fatigue testing of Ni nanowires based on a micro-mechanical device. Exp Mech. 2017;57(3):495-500. https://doi.org/10.1007/s11340-016-0199-1

[13] Akbarian S, Dehghani K. On the molecular dynamics simulation of fatigue behavior of pre-cracked aluminum chip for NEMS application: Effect of cyclic loading mode and surface roughness geometry. Int J Fatigue. 2020;105570. https://doi.org/10.1016/j.ijfatigue.2020.105570

[14] Potirniche GP, Horstemeyer MF, Jelinek B, et al. Fatigue damage in nickel and copper single crystals at nanoscale. Int J Fatigue. 2005;27(10-12):1179-1185. https://doi.org/10.1016/j.ijfatigue.2005.06.015

[15] Chen B, Wu W-P, Chen M-X, et al. Molecular dynamics study of fatigue mechanical properties and microstructural evolution of Ni-based single crystal superalloys under cyclic loading. Comput Mater Sci. 2020;185:109954. https://doi.org/10.1016/j.commatsci.2020.109954

[16] Potirniche GP, Horstemeyer MF, Wagner GJ, et al. A molecular dynamics study of void growth and coalescence in single crystal nickel. Int J Plast. 2006;22(2):257-278. https://doi.org/10.1016/j.ijplas.2005.02.001

[17] Saini K, Kumar N. Torsional deformation behavior of cracked gold nano-wires. Acta Mech. 2014;225(3):687-700. https://doi.org/10.1007/s00707-013-0993-0

[18] PLIMPTON S. LAMMPS: molecular dynamics simulator. http://lammps.sandia.gov/. lammps sandia gov. 2018;

[19] Mendelev MI, Kramer MJ, Hao SG, et al. Development of interatomic potentials appropriate for simulation of liquid and glass properties of nizr2 alloy. Philos Mag. 2012;92(35):4454-4469. https://doi.org/10.1080/14786435.2012.712220

[20] Rassoulinejad-Mousavi SM, Mao Y, Zhang Y. Evaluation of copper, aluminum, and nickel interatomic potentials on predicting the elastic properties. J Appl Phys. 2016;119(24). https://doi.org/10.1063/1.4953676

[21] Wen Y-H, Zhu Z-Z, Zhu R-Z. Molecular dynamics study of the mechanical behavior of nickel nanowire: Strain rate effects. Comput Mater Sci. 2008;41(4):553-560. https://doi.org/10.1016/j.commatsci.2007.05.012

[22] Liu Y, Zhao J. The size dependence of the mechanical properties and breaking behavior of metallic nanowires: A statistical description. Comput Mater Sci. 2011;50(4):1418-1424. https://doi.org/10.1016/j.commatsci.2010.11.026

[23] Hoover WG. Canonical dynamics: equilibrium phase-space distributions. Phys Rev A. 1985;31(3):1695. https://doi.org/10.1103/PhysRevA.31.1695

[24] Stukowski A. Visualization and analysis of atomistic simulation data with OVITO-the Open Visualization Tool. Model Simul Mater Sci Eng. 2010;18(1):15012. https://doi.org/10.1088/0965-0393/18/1/015012

[25] Stukowski A, Bulatov V V, Arsenlis A. Automated identification and indexing of dislocations in crystal interfaces. Model Simul Mater Sci Eng. 2012;20(8):85007. https://doi.org/10.1088/0965-0393/20/8/085007

[26] Neogi A, Mitra N. Shock induced deformation response of single crystal copper: Effect of crystallographic orientation. Comput Mater Sci. 2017;135:141-151. https://doi.org/10.1016/j.commatsci.2017.04.009

[27] Diao J, Gall K, Dunn ML, et al. Atomistic simulations of the yielding of gold nanowires. Acta Mater. 2006;54(3):643-653. https://doi.org/10.1016/j.actamat.2005.10.008

[28] Gupta P, Katakam KC, Katakareddi G, et al. Crack and its interaction with defects in Al coated with Cu 50 Zr 50 metallic glass thin film: an MD simulation study. J Mol Model. 2020;26(4):1-17. https://doi.org/10.1007/s00894-020-4335-6

[29] Rupert TJ. Strain localization in a nanocrystalline metal: Atomic mechanisms and the effect of testing conditions. J Appl Phys. 2013;114(3):33527. https://doi.org/10.1063/1.4815965

[30] Park HS, Gall K, Zimmerman J a. Deformation of FCC nanowires by twinning and slip. J Mech Phys Solids. 2006;54(9):1862-1881. https://doi.org/10.1016/j.jmps.2006.03.006

[31] Wang P, Chou W, Nie A, et al. Molecular dynamics simulation on deformation mechanisms in body-centered-cubic molybdenum nanowires. J Appl Phys. 2011;110(9):93521. https://doi.org/10.1063/1.3660251

[32] Ahn DC, Sofronis P, Kumar M, et al. Void growth by dislocation-loop emission. J Appl Phys. 2007;101(6):63514. https://doi.org/10.1063/1.2710346

[33] Pohjonen AS, Djurabekova F, Nordlund K, et al. Dislocation nucleation from near surface void under static tensile stress in Cu. J Appl Phys. 2011;110(2):23509. https://doi.org/10.1063/1.3606582

[34] Traiviratana S, Bringa EM, Benson DJ, et al. Void growth in metals: Atomistic calculations. Acta Mater. 2008;56(15):3874-3886. https://doi.org/10.1016/j.actamat.2008.03.047

[35] Bobylev S V, Gutkin MY, Ovid'ko IA. Partial and split dislocation configurations in nanocrystalline metals. Phys Rev B. 2006;73(6):64102. https://doi.org/10.1103/PhysRevB.73.064102

[36] Liu P, Wang L, Yue Y, et al. Room-temperature superplasticity in Au nanowires and their atomistic mechanisms. Nanoscale. 2019;11(18):8727-8735. https://doi.org/10.1039/C9NR02053E

[37] Wang BZ, Liu DS, Wen ZX, et al. Tension/compression asymmetry of [001] single-crystal nickel-based superalloy DD6 during low cycle fatigue. Mater Sci Eng A. 2014;593:31-37. https://doi.org/10.1016/j.msea.2013.09.013

[38] Jabbari AH, Sedighi M, Jahed H, et al. Low cycle fatigue behavior of AZ31B extrusion at elevated temperatures. Int J Fatigue. 2020;139:105803. https://doi.org/10.1016/j.ijfatigue.2020.105803

[39] Zhang ZF, Wang ZG, Sun ZM. Evolution and microstructural characteristics of deformation bands in fatigued copper single crystals. Acta Mater. 2001;49(15):2875-2886. https://doi.org/10.1016/S1359-6454(01)00219-1

[40] Buque C. Dislocation structures and cyclic behaviour of [011] and [111]-oriented nickel single crystals. Int J Fatigue. 2001;23(8):671-678. https://doi.org/10.1016/S0142-1123(01)00032-9

[41] Brammer AT, Jordon JB, Allison PG, et al. Strain-controlled low-cycle fatigue properties of extruded 6061-T6 aluminum alloy. J Mater Eng Perform. 2013;22(5):1348-1350. https://doi.org/10.1007/s11665-012-0411-0

[42] Luo J, Dahmen K, Liaw PK, et al. Low-cycle fatigue of metallic glass nanowires. Acta Mater. 2015;87:225-232. https://doi.org/10.1016/j.actamat.2014.12.038

[43] Sha Z, Wong WH, Pei Q, et al. Atomistic origin of size effects in fatigue behavior of metallic glasses. J Mech Phys Solids. 2017;104:84-95. https://doi.org/10.1016/j.jmps.2017.04.005

[44] Veerababu J, Goyal S, Sandhya R, et al. Understanding the Low Cycle Fatigue Behavior of Single Crystal Cu at the Nano-scale: A Molecular Dynamics Study. Trans Indian Inst Met. 2017;70(3):867-874. https://doi.org/10.1007/s12666-017-1066-1

Microstructure evolution during high cycle fatigue in a low carbon micro-alloyed dual-phase steel sheet

Gaurav Pandey[1,a*], Basudev Bhattacharya[2,b], Somjeet Biswas[1,c*]

[1]Light Metals and Alloys Research Lab, Department of Metallurgical and Materials Engineering, Indian Institute of Technology, Kharagpur, West Bengal, 721302, India

[2]R and D, Tata Steel, Jamshedpur, Jharkhand, 831001, India

[a]gaurav01340@gmail.com, [b]drbasudevb@gmail.com, [c]somjeetbiswas@metal.iitkgp.ac.in

Keywords: Dual Phase (DP) Steel, High Cycle Fatigue (HCF), Microstructure Evolution, EBSD, Crystallographic Texture, Damage Mechanism

Abstract. Ferrite–martensite dual-phase (FMDP) steel sheets processed by cold rolling followed by intercritical annealing and quenching were subjected to high-cycle fatigue (HCF) testing to investigate fatigue-induced microstructural evolution. Fatigue tests were conducted at stress amplitudes of 485–580 MPa, and the resulting microstructural changes were examined using SEM and EBSD. The specimen tested at 485 MPa, corresponding to the endurance limit, exhibited a stable microstructure with negligible plastic strain localization. In contrast, higher stress amplitudes led to the formation of persistent slip bands, microcrack initiation at ferrite–martensite interfaces, grain refinement, and significant increases in crystallographic misorientation. EBSD-based GOS, LAM, and KAM analyses revealed enhanced dislocation accumulation and localized plastic deformation at higher stress levels, correlating well with accelerated fatigue damage and premature failure.

Introduction

The automotive industry increasingly uses thin-gauge Advanced High-Strength Steel (AHSS) sheets to reduce vehicle weight, CO_2 emissions, and cost while improving fuel efficiency and safety [1]. Since the 1970s, AHSS has evolved through three generations with enhanced strength, ductility, and fatigue resistance [1]. First-generation AHSS improved strength at low cost but exhibited limited damage tolerance due to ferrite–martensite decohesion and martensite cracking [1,2]. Second-generation AHSS offered superior strength–ductility balance but suffered from segregation, poor weldability, complex processing, and high alloying costs [2]. These limitations led to the development of third-generation AHSS, which provides improved mechanical performance at reduced cost [1,2,3]. Among these, ferrite–martensite dual-phase (FMDP) steels are particularly attractive due to their high strength, good ductility, continuous yielding behaviour, and superior fatigue resistance [4].

FMDP steels primarily consist of ferrite and martensite, with minor retained austenite and bainite depending on processing [4]. Strength is governed by martensite fraction, ferrite grain refinement [3], and ferrite dislocation density [5], while ductility is enhanced through optimized martensite morphology [5], ferrite grain refinement [3], CSL-type ferrite–martensite interfaces formed via K–S transformation [4,5], texture weakening, and TRIP-assisted deformation from stabilized retained austenite [6]; formability is further improved by increasing normal anisotropy and reducing planar anisotropy through tailored TMP routes [4,5]. Notably, Ashrafi et al. achieved 734 MPa UTS with 42% ductility in 0.18 wt.% C DP steel via IA&Q [7], while Singh et al. reported 544 MPa strength with uniform and fracture ductility of 32% and 49%, respectively, in 0.05 wt.% C DP steel processed by CR+Q&P, attributed to ~33% island-type martensite, fine ferrite grains, high dislocation density, increased geometrically necessary boundaries, and enhanced CSL boundary fractions [4].

FMDP steel sheets are widely used in thin-gauge automotive components such as B-pillars, roof and door panels, side panels, cowl panels, and floor panels [1]. These components experience severe dynamic loading during service, which can induce microstructural degradation and premature failure, making high-cycle fatigue (HCF) assessment essential [8]. Previous studies show that fatigue performance strongly depends on martensite fraction and morphology: fatigue life increases up to ~30% martensite but deteriorates at higher fractions due to ferrite–martensite decohesion and martensite cracking [9], with inferior cyclic performance reported at $V_m \approx 0.60$ compared to $V_m \approx 0.45$ [10], and superior endurance in fibrous martensite over network morphology [11]. Despite these findings, quantitative understanding of microstructural evolution during HCF in ferritic–martensitic DP steel sheets remains limited. In this study, lean-alloy (~0.05 wt.% C) FMDP steel sheets were developed and subjected to HCF testing, with quantitative microstructural and texture evolution analysed using SEM and EBSD to correlate fatigue endurance with microstructural changes.

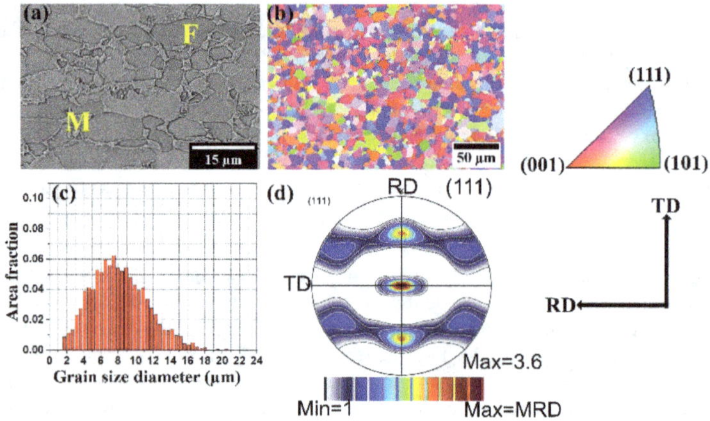

Fig. 1 (a) SEM microstructure, (b) inverse pole figure map (IPF), (c) grain size distribution, and (d) pole figure of the low carbon micro-alloyed dual phase steel

Experimental procedure

A hot-rolled low-carbon micro-alloyed steel sheet (0.05C–1.3Mn–0.6Cr–0.4Si), ~6.7 mm thick) with an initial ferrite–0.45 bainite microstructure was developed. The austenite formation temperatures were Ac_1 = 680 °C and Ac_3 = 880 °C, measured using a BAHR DIL 805 A/D dilatometer. The sheet was normalized at 920 °C for 5 min, followed by ~84% cold rolling (von Mises strain ~2.1) in five passes. Intercritical annealing was performed at 790 ± 20 °C for 5 min, followed by water quenching at ~200 °C/s to obtain a ferrite–0.25 martensite dual-phase microstructure. High-cycle fatigue (HCF) specimens were prepared according to ASTM E466 and tested at room temperature (25 °C) using an Instron Electropulse E10000 machine at a stress ratio of R = 0 and a frequency of 30 Hz. Fatigue tests were conducted at stress amplitudes of 485 MPa (specimen A) 520 MPa (specimen B), 550 MPa (specimen C), and 580 MPa (specimen D). Post-fatigue quantitative microstructural analyses were subsequently performed. Microstructural characterization was carried out using secondary electron imaging and EBSD on a Zeiss Sigma 300 VP field-emission SEM. For SE imaging, RD–TD sections were polished and etched with 1%

Nital for 15 s. EBSD samples were mirror-polished using SiC papers (up to 4000 grit), diamond suspension (~1 μm), and colloidal silica (0.03 μm), followed by electropolishing in a 1:4 perchloric acid–ethanol solution using a Struers Electropol 5 at 20 kV for 20 s. SEM images were analyzed using ImageJ, while EBSD data were processed using TSL-OIM software.

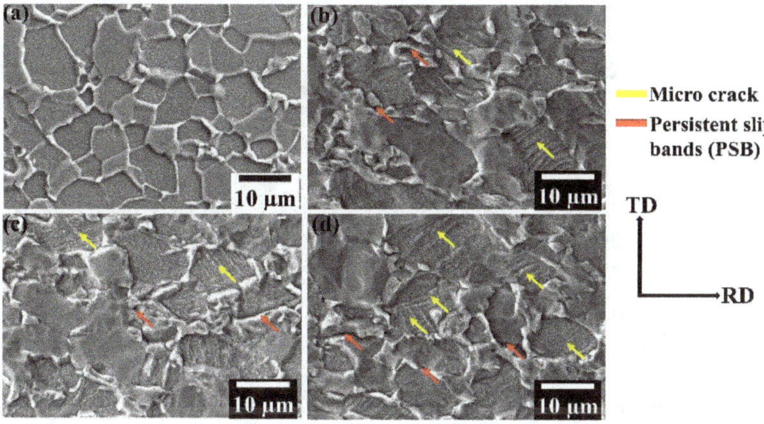

Fig. 2 SEM microstructures of the specimens tested at (a) 485, (b) 520, (c) 550 MPa, and (d) 580 MPa stress amplitude

Results and Analysis

The SEM image, inverse pole figure (IPF) map, grain size distribution, and pole figure of the DP steel sheet are shown in Fig. 1(a–d), respectively. The initial microstructure consists of ~25% island-type martensite uniformly distributed within the ferrite matrix (Fig. 1a), where ferrite appears darker and martensite slightly brighter in SEM images. The average ferrite grain size and martensite island size were ~8 μm and ~2 μm, respectively, while the EBSD-derived grain size distribution ranged from 2 to 25 μm with a mean of ~8 μm (Fig. 1c). The IPF map (Fig. 1b) shows grain orientations based on the standard crystallographic color key, though ferrite and martensite could not be distinctly resolved. The pole figure (Fig. 1d) reveals a strong γ-fibre texture (111)||ND with an intensity of ~3.6 MRD. SEM microstructures of specimens A–D after HCF testing are shown in Fig. 2(a–d). The martensite morphology remained unchanged across all stress amplitudes; however, specimens B, C, and D tested at 520, 550, and 580 MPa exhibited microcracks along ferrite–martensite interfaces accompanied by persistent slip bands (PSBs) in ferrite grains (Fig. 2b–d). Microcracks and PSBs are indicated by red and yellow arrows, respectively. The EBSD-derived area fraction grain size distributions considering RHAGBs (15°–65°) are presented in Fig. 3. A progressive shift of the distribution toward smaller grain sizes from specimen A to D indicates grain refinement with increasing stress amplitude. Figure 4 shows the grain orientation spread (GOS), local average misorientation (LAM), and kernel average misorientation (KAM) maps for the initial material and specimens A–D tested at 485, 520, 550, and 580 MPa. The initial material exhibited low average GOS, LAM, and KAM values of 0.57°, 0.49°, and 0.30°, respectively (Fig. 4a). These values slightly decreased at 485 MPa but increased markedly at higher stress amplitudes, reaching average GOS values of 3.5°, 3.7°, and 4.2°, LAM values of 0.92°, 1.01°, and 1.23°, and KAM values of 0.75°, 0.78°, and 0.80° for specimens tested at 520, 550, and 580 MPa, respectively (Fig. 4c–e), indicating enhanced lattice distortion and strain accumulation.

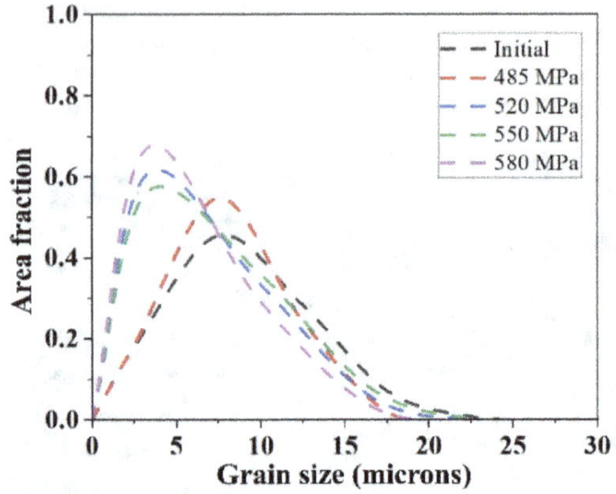

Fig.3. Grain size distribution obtained from the EBSD analysis of the initial specimen and specimens tested at (a) 485 MPa, (b) 520 MPa, (c) 550 MPa, and (d) 580 MPa stress amplitude

Discussion

Following cold rolling and intercritical annealing with quenching (IA&Q) at 790 ± 20 °C for 5 min, a ferrite–martensite dual-phase (FMDP) steel sheet was obtained and subjected to high-cycle fatigue (HCF) testing at stress amplitudes of 485, 520, 550, and 580 MPa to examine fatigue-induced microstructural evolution. At a stress amplitude of 485 MPa (specimen A), corresponding to the endurance limit, the microstructure remained comparable to the initial condition and showed no evidence of low-energy dislocation configurations such as substructures, channels, ladders, or slip bands (Fig. 1a). This behaviour indicates negligible lattice distortion, limited grain rotation, and minimal plastic strain localization, enabling the material to withstand a large number of loading cycles without failure while maintaining a stable microstructure. In contrast, specimens tested at higher stress amplitudes (520–580 MPa) exhibited pronounced microstructural degradation. These changes included the formation of dislocation substructures, ladder-like arrangements, persistent slip bands (PSBs), and microcrack initiation sites within the ferrite phase (Fig. 2a–c). The density of PSBs increased systematically with increasing stress amplitude, reflecting enhanced cyclic plastic strain localization in the ferritic matrix. Fatigue cracks preferentially nucleated along PSBs inclined at approximately 45° to the loading direction, corresponding to the orientation of maximum resolved shear stress on active slip systems, ultimately leading to failure [12]. The fatigue damage process followed the classical four-stage sequence: (i) crack initiation, (ii) stage-I crack growth along slip bands governed by shear stress, (iii) stage-II crack propagation perpendicular to the maximum tensile stress, and (iv) final ductile fracture upon reaching a critical crack length [13]. In addition, specimens tested at 520, 550, and 580 MPa exhibited a noticeable reduction in grain size, which is attributed to fatigue-induced dynamic recrystallization (DRX) under severe cyclic deformation conditions [14]. To quantitatively assess fatigue-induced microstructural evolution, EBSD-based grain orientation spread (GOS), local average misorientation (LAM), and kernel average misorientation (KAM) analyses were employed. These metrics effectively capture localized plastic strain accumulation associated with lattice distortion and grain rotation during cyclic loading. In particular, LAM and KAM provide insights into intragranular strain heterogeneity, while KAM reflects local misorientation between neighbouring pixels and is closely linked to dislocation density [8]. Both the initial material and the specimen tested at 485 MPa exhibited low misorientation levels, with

LAM and KAM distributions concentrated near the origin (Fig. 4a–b), indicating minimal dislocation accumulation and negligible plastic strain localization. Quantitatively, the initial material showed average LAM and KAM values of 0.49° and 0.30°, which slightly decreased to 0.45° and 0.26°, respectively, for the specimen tested at 485 MPa. This observation is consistent with the absence of PSBs, substructures, or microcracks at this stress amplitude, confirming that 485 MPa represents the endurance limit where the microstructure remains largely stable under cyclic loading. In contrast, specimens subjected to higher stress amplitudes of 520, 550, and 580 MPa displayed a pronounced shift of LAM and KAM distributions toward higher misorientation values (Fig. 4a–b), reflecting increased dislocation accumulation and severe plastic strain localization. The average LAM values increased to 0.92°, 1.01°, and 1.23°, while the corresponding KAM values rose to 0.75°, 0.78°, and 0.80°, respectively. These trends are consistent with the observed formation of PSBs and microcrack initiation at ferrite–martensite interfaces, as discussed earlier. Overall, the combined GOS, LAM, and KAM analyses clearly distinguish the fatigue response below and above the endurance limit. While the specimen tested at 485 MPa maintains low lattice distortion and stable microstructural characteristics, higher stress amplitudes promote significant crystallographic misorientation, localized plastic strain accumulation, and accelerated fatigue damage, ultimately resulting in premature failure at lower numbers of cycles.

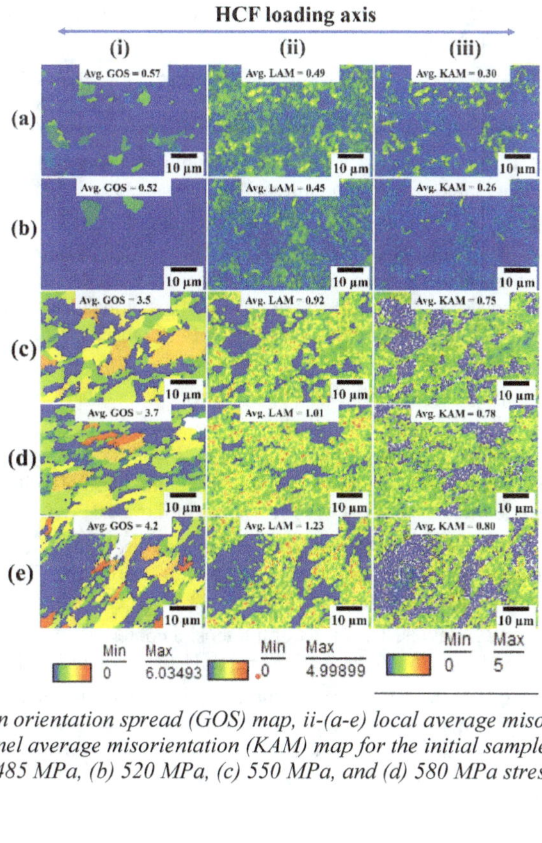

Fig. 4 i-(a-e) Grain orientation spread (GOS) map, ii-(a-e) local average misorientation (LAM) map iii-(a-e) kernel average misorientation (KAM) map for the initial sample and specimens tested at (a) 485 MPa, (b) 520 MPa, (c) 550 MPa, and (d) 580 MPa stress amplitude

Conclusion

(a). FMDP steel tested at 485 MPa (endurance limit) exhibited a stable microstructure with minimal lattice distortion, low misorientation, and absence of persistent slip bands or microcracks. (b). Increasing stress amplitude (520–580 MPa) resulted in pronounced fatigue-induced microstructural degradation, including PSB formation, microcrack initiation, and grain refinement. (c). EBSD-based GOS, LAM, and KAM analyses quantitatively demonstrated enhanced dislocation accumulation and localized plastic strain at higher stress amplitudes. (d). The combined microstructural and crystallographic evidence clearly distinguishes fatigue behaviour below and above the endurance limit, providing insight into fatigue damage mechanisms in FMDP steel sheets.

References

[1] A.H. Steels, September 2016 Application Guidelines, (2016).

[2] Y. Mazaheri, A.H. Jahanara, M. Sheikhi, A.G. Kalashami, High strength-elongation balance in ultrafine grained ferrite-martensite dual phase steels developed by thermomechanical processing, Materials Science and Engineering: A 761 (2019) 138021. https://doi.org/10.1016/j.msea.2019.06.031.

[3] Y. Mazaheri, A. Kermanpur, A. Najafizadeh, A novel route for development of ultrahigh strength dual phase steels, Materials Science and Engineering: A 619 (2014) 1–11. https://doi.org/10.1016/j.msea.2014.09.058.

[4] A.K. Singh, D.K. Chouhan, B. Bhattacharya, S. Biswas, High strength-ductility combination by quenching and partitioning of a low carbon microalloyed dual-phase steel, Materials Science and Engineering: A 870 (2023) 144854. https://doi.org/10.1016/j.msea.2023.144854.

[5] A.K. Singh, B. Bhattacharya, S. Biswas, High strength, ductility and sheet formability by normalizing and quenching of low carbon microalloyed dual-phase steel, Materials Science and Engineering: A 890 (2024) 145848. https://doi.org/10.1016/j.msea.2023.145848.

[6] S. Yan, X. Liu, W.J. Liu, T. Liang, B. Zhang, L. Liu, Y. Zhao, Comparative study on microstructure and mechanical properties of a C-Mn-Si steel treated by quenching and partitioning (Q&P) processes after a full and intercritical austenitization, Materials Science and Engineering A 684 (2017) 261–269. https://doi.org/10.1016/j.msea.2016.12.026.

[7] H. Ashrafi, M. Shamanian, R. Emadi, N. Saeidi, A novel and simple technique for development of dual-phase steels with excellent ductility, Mater. Sci. Eng. A 680 (2017) 197–202, https://doi.org/10.1016/j.msea.2016.10.098.

[7] G.T. Zhang, N.Q. Zhu, B.W. Sun, Z.Z. Zhao, Z.W. Zheng, D. Tang, L. Li, Effect of v addition on microstructure and mechanical properties in c–mn–si steels after quenching and partitioning processes, Metals (Basel) 11 (2021). https://doi.org/10.3390/met11081306.

[8] P. Ghosal, A. Raj, S.K. Paul, Influence of uniaxial and biaxial pre-straining on the low cycle fatigue performance of DP590 steel, Int. J. Fatigue 149 (2021) 106260, https://doi.org/10.1016/j.ijfatigue.2021.106260.

[9] A.M. Sherman, R.G. Davies, The effect of martensite content on the fatigue of a dual-phase steel, Int. J. Fatigue 3 (1981) 36–40, https://doi.org/10.1016/0142- 1123(81)90047-5.

[10] P.C. Chakraborti, M.K. Mitra, Microstructural response on the room temperature low cycle fatigue behaviour of two high strength duplex ferrite–martensite steels and a normalized ferrite–pearlite steel, Int. J. Fatigue 28 (2006) 194–202, https:// doi.org/10.1016/j.ijfatigue.2005.06.024.

[11] M.J. Molaei, A. Ekrami, The effect of dynamic strain aging on fatigue properties of dual phase steels with different martensite morphology, Mater. Sci. Eng. A 527 (2009) 235–238, https://doi.org/10.1016/j.msea.2009.08.005.

[12] G.E. Dieter, D. Bacon, Mechanical Metallurgy 3, McGraw-Hill, New York, 1976, pp. 43–53.

[13] Y. Dai, S. Wang, Q. He, C. Liu, X. Wang, X. Li, L. Li, Y. Liu, C. He, Q. Wang, Effect of microstructure on slip-induced crack initiation and early propagation of martensitic steel during high cycle fatigue, Int. J. Fatigue 167 (2023) 107275, https://doi.org/10.1016/j.ijfatigue.2023.107275.

[14] D. Yu, D. Zhang, Y. Luo, J. Sun, J. Xu, F. Pan, Microstructure evolution during high cycle fatigue in Mg-6Zn-1Mn alloy, Mater. Sci. Eng. A 658 (2016) 99–108, https://doi.org/10.1016/j.msea.2016.01.061.

Effect of chemical treatment on the thermo-mechanical properties of the natural fiber-reinforced composites

Alok PATEL[1,a], Sanjit BISWAS[1,b*], Sarada Prasad PARIDA[2,c*]

[1]Department of Civil Engineering, NIT Warangal, Telangana, India

[2]Konark Institute of Science and Technology, Khordha, Odisha, India 752050

[a]ap21cerep09@student.nitw.ac.in, [b]sbiswas@nitw.ac.in, [c]sarada800@gmail.com

Keywords: Chemical Processing, Alkali treatment, Na-OH, Epoxy, Rice-Straw

Abstract. Nowadays, the management of agricultural waste is a major challenge for authorities and its burning is still uncontrolled that causes severe health issues. In other hand, shortfall in conventional wood awakened researchers to fabricate new composite variants such as natural fiber reinforced polymer composites. However, the use of natural fibers could not incur sufficient mechanical and thermal strengths in the composites. In this work, to demonstrate the sustainable change in thermal and mechanical properties in composites by chemical treatment; alkali treatments of the rice straw and husk are carried out. Hybrid-composite specimens are manufactured by hand layup process, mechanical (hardness, impact, tensile, and flexural) and thermal tests (TGA/DTG, flammability, and thermal absorption) are conducted. Comparative studies show that about a 30-40% rise in mechanical strength and thermal stability is achieved by alkali treatment advocated by SEM finding that shows alkali treatment enables superior fibre-matrix bonding.

Introduction

The burning of agricultural waste in India is becoming a serious issue. Around 98MT of crop wastes (bast, leaf, or stem of sugar cane, wheat, rice, and other crops) are burnt annually [1]. According to the World Bank, in developing countries such as India, the urbanization rate has increased and will grow to 40.76% by 2030. Hence, a huge demand for infrastructure built-up materials will be needed. Meanwhile, the increase in urbanization and other environmental factors leads to a decrease in the forest area [2] and thus it cannot fulfill the rising demand. In this aspect, researchers are proposing composites from natural fiber (NF) and agricultural waste [3,4]. The degradation of these do not leave poisonous substances [5] with fewer environmental issues [6] reduced energy demand [7] and carbon footprints [8,9]. Furthermore, natural fiber-reinforced composites have a lower cost of production with an estimation of US $200–$1000 and 4 GJ energy requirements for one ton [10]. Amid such advantages, natural fiber reinforced polymeric composites (NFRPC) have the least use for less mechanical strength. However, this limitation can be eliminated after the chemical processing of NF [11,12].

From the literature survey, it is observed that the chemical processing of NF improves both mechanical and thermal characteristics of NFRPC. Only a very few work citing the use of rice straw as a reinforcing agent. From Indian context there is huge scope to prepare rice husk and straw reinforced polymeric composites. This will not only fulfil the growing demand of the material but also deal with the management of agriculture waste and pollution control. Here, NFRPCs are prepared from rice straw(RS) and rice husk(RH), in epoxy by hand layup process. Mechanical and thermal analysis have been conducted and results of alkali treated NFRPCs are compared with the properties of NFRPCs prepared from untreated fibre.

Natural fiber and chemical processing

A fiber is the major structural part of a composite that provides strength, shape, and rigidity. The plant fibers may be extracted from the bast, seed, leaf, fruit, stalk, wood, or reed. It comprised of cellulose, hemicelluloses, lignin, pectin, natural resin, and water-soluble minerals[13]. The rice straw and rice husk are primarily composed of 30-40% cellulose. Apart from this, 15-20% of hemicelluloses and 20-25% lignin is present. Typically the rice straw is comprised of a bundle of cellulosic fiber called microfibrils having a diameter of 3-4nm. The microfibrils are bonded together by hemicelluloses and lignin. The untreated fibre contains hydroxyl groups as a constituent of lignin and thus swells by acquiring moisture as a result crack formation occurs. The wax layer found around the fibers leads to poor compatibility with the matrix for bonding leading to lower mechanical strength, thermal stability and Poor durability. As a result, a plant fiber is least preferred [14]. There are various methods of chemical treatment of natural fibers However, the alkali treatment is opted for its versatility and easy availability [15]. In the alkali treatment process, sodium hydroxide (NaOH) is diluted in the de-mineralized water by weight percentage of 1-2% to form an alkali solution. Then fiber to be treated is soaked and left for a few hours (usually 4 to 6 hrs). Then it is rinsed well in the distilled water to rinse the leftover components of the alkali. The chemical reaction associated with the alkali treatment is given by;

$$Fibre-cell-OH+NaOH \rightarrow Fibre\ cell-ONa^{+}+H_2O+impurities$$

Material Preparation

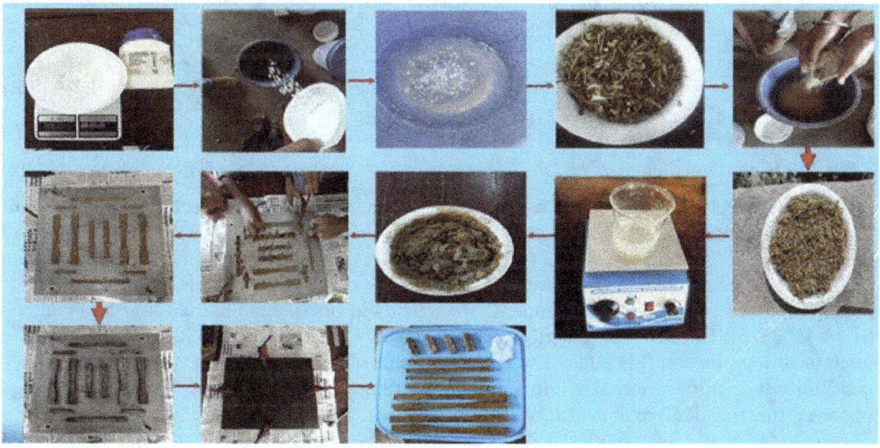

Fig. 1. Process of preparation of test samples.

In the current study, RS is collected and chopped into tiny pieces about 2-3mm size. The RH (5-7 μm) along with RS is processed with an alkali solution (2% of NaOH). After alkali treatment, the rice husk and husk are thoroughly washed in distilled water and sundried for about 48 hours. A mold (Fig. 2) made of stailess steel used for preparation of tensile, flexural, and impact test specimens. The mold was cleaned and the polish wax was applied thoroughly for quick removal of the specimen. The base is prepared manually by mixing epoxy (LY556) with its reagent (MS91) in a ratio of 10:1. Then the fiber is added to it. The weight percentage of fiber to the base is kept 50:50. Here, the fiber is comprised of an equal amount of RS & RH. The mixture was then poured into the mold cavity for preparation of the test specimens. The mold is tightened and left for 24 hours of curing and then the specimens are released from the mold.

Material Testing

Five test specimens each of size 165×19×5 mm are prepareed for tensile from untreated(UT) and alkali-treated(AT) fibers according to ASTM D3039 performed in INSTRON 3382 universal testing machine available at VSSUT, Burla with a low feed rate of 2mm/min. In same machine, the flexural test (ASTM D790) with 1mm/min is conducted and the flexural strength is measured. Also, Vickers hardness test (ASTM D785) and impact test (ASTM D256) is conducted. Further, the thermal stability is analyzed through thermoglavometrtic analysis following ASTM E1131 stndard for 25mg samples inside the closed chamber of nitrogen maintaining a flow of 75mL/min from 30°C to 800°C with constant temperature rise of 10°C/min. The weight loss with respect to temperature and time is reported through graph. Further, the thermal absorptivity in terms of rate of cooling and heating of the specimens are tested under direct sunlight and burning test is performed.

Result discussion

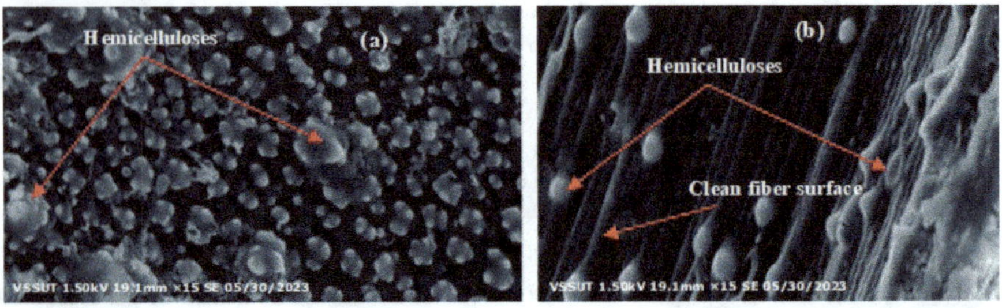

***Fig. 2,** Microscopic image of surface of rice straw; (a) untreated; (b) treated with NaOH (2%)*

The alkali treatments in the fiber break the highly packed cellulose present in the fiber which is crystalline and breaks the chain of cellulose and makes it amorphous. Thus a bunch of tiny fibers are formed. Fig. 2(a) and Fig. 2(b) show the SEM images of the fiber surface of an untreated fiber (UTF) and the alkali treated fiber (ATF). the fiber fully covered with hemicellulose and lignin is mostly removed after the alkali treatment. This results in fiber length to diameter ratio reduction by 14-15% [16]. This hair like structure formed on the fiber surface interlock with the resin during manufacturing process. It is a fact that, the more number of hair like structure is formed; there is possibility to facilitate more reaction site during curing process of matrix. From the study, it is observed that the alkali treatment increases the tensile strength of composites by 32% (Fig. 3(a)) due to the fact that, more number of ATFs are need to be reinforced in the specimen as compared to UTFs to maintain constant fiber volume fraction. Further the hair like structure formed on ATFs facilitates interfacial adhesion of fiber and matrix so that load transfer between matrix and fiber occurs smoothly enabling to carry higher tensile loads [17].

Similar to tensile strength, ATF reinforced composites found to have 36.4% higher flexural strength over UTF reinforced composited as compared in Fig. 4(b) due to the modification of fiber by alkali treatment. The impact strength, and hardness of the ATF reinforced composite is found to be increased by 19.78% and 28.88% respectively as shown by Fig. 4 (c &d). The thermal stability of the samples is compared through TGA and DTG analysis. It is observed that both the samples show the same nature of material decomposition to heat, however, the initial stability of composites with a difference of 10°C is observed with the ATF reinforced composites (Fig. 5). It is due to the fact that processing of the natural fiber with alkali breaks the cellulose chain and

Materials Research Proceedings 65 (2026) 58-63 https://doi.org/10.21741/9781644904138-9

removes low molar mass compound from fiber. As a result enables to sustain on high initial thermal decomposition [18]. On further thermal decomposition, the chemically processed fiber-reinforced composite shows a little lower rate of thermal decomposition. The thermal absorptivity of the prepared specimens are tested when exposed to direct sunlight. For the process, the two test samples are kept under sunlight where the ambient temperature was observed to be 21.8°C. The experiment is conducted in two field where, the samples were first exposed to sunlight and the rise of surface temperature is measured till it becomes uniform. Then the samples were brought inside the room and the rate of cooling is determined through reading of the surface temperature as shown in Fig. 6(a). It is observed that the UTF composites exhibits higher surface temperature than ATF composites both in heating and cooling implying the higher rate of thermal absorptivity. Further, the flammability of the prepared samples are tested and expressed as the ignition time and temperature (Fig. 6(b)). It is observed that the ignition time increases from 21.92 Sec to 27.77 Sec and ignition temperature increases to 336°C from 300°C due to the alkali treatment of the natural fibers. This is possibly due to the loss of lignin contents from the natural fibers by alkali treatment that enables to bond with the epoxy to mora a more compact bonding during manufacturing.

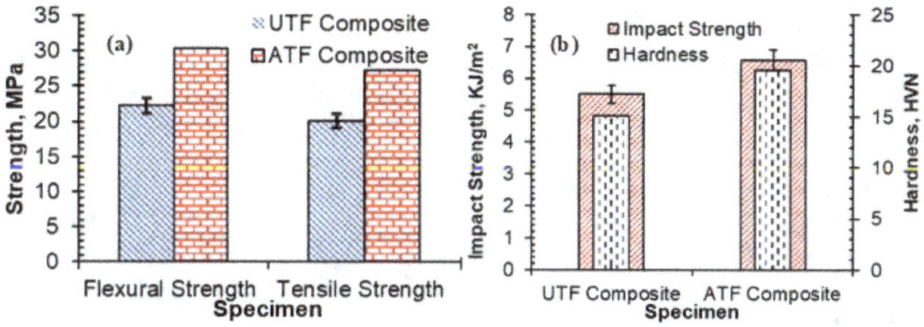

Fig 4, Comparison of Mechanical Strength (a) tensile & Flexural; (b) Impact & Hardness

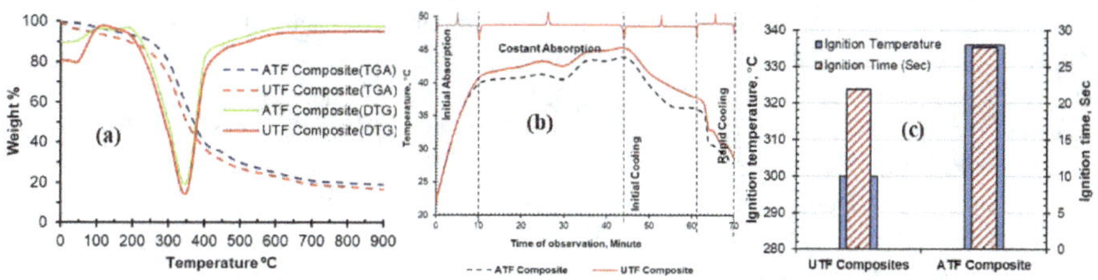

Fig 5, Comparison of Thermal stability (a) TGA /DTG (b) Thermal absorptivity (b) Ignition time and temperature

Conclusion

Following conclusions are observed from this study;

- The chemical processing breaks the cellulose and hemicelluloses present in natural fiber so that a bunch of tiny fibers are formed in place of a single stack of fiber observed through SEM study. The aspect ratio of fiber is reduced and the tiny hair like structures facilitates inter facial bonding between fiber and matrix.

- Composites prepared from the alkali-treated fiber and epoxy show an overall 30-40% rise in tensile, flexural, impact strength, and hardness of composites reinforced with alkali treated fibre as composites fabricated by reinforcing untreated fibers.
- The initial thermal stability of the composites is increased by 10°C by the chemical processing of reinforced fibers.

References

[1] K. Kaur, P. Singh, Crop Residue Burning in India: Potential Solutions. Agricultural Waste - New Insights 2022. http://dx.doi.org/10.5772/intechopen.107457.

[2] Gore, Al, *Our Choice: A Plan to Solve the Climate Crisis*. Rod ale Books. pp. 174-192, 2009. ISBN 978-1-59486-734-7.

[3] M. Zwawi, A review on natural fiber bio-composites, surface modifications and applications. Molecules. 26(2021)404.

[4] X. Liu, E. Zhang, Z. Feng. Degradable bio-based epoxy vitrimers based on imine chemistry and their application in recyclable carbon fiber composites. *J Mater Sci* **56**(2021)15733–15751 https://doi.org/10.1007/s10853-021-06291-5.

[5] M. Khemakhem, K. Lamnawar, A. Maazouz, M. Jaziri, Biocomposites based on polylactic acid and olive solid waste fillers: Effect of two compatibilization approaches on the physicochemical, rheological, and mechanical properties. Polymer Composites. 39(2018)E152-63.

[6] S. Ojha, G. Raghavendra, S.K. Acharya, A Comparative Investigation of Bio Waste Filler (Wood Apple-Coconut) Reinforced Polymer Composites. Polym. Compos. 35(2014)180–185.

[7] R. Rahman, M. Rahman, S. Hamdan, F. Hossen, J.C. Lai, F.K. Liew, Effect of nanoclay and silica on mechanical and morphological properties of jute cellulose polyethylene biocomposites. Journal of Vinyl and Additive Technology. 23(2017)113-8.

[8] S.H. Bhosale, V.V. Singh, M.C. Rangasai, S. Bandyopadhyay-Ghosh, S.B. Ghosh, Integration of biobased functionalized feedstock and plastisol in epoxy resin matrix toward developing structural jute biocomposites with enhanced impact strength and moisture resistance properties. Polymer Composites. 37(2016)391-7.

[9] A. Ramzy, D. Beermann, L. Steuernagel, D. Meiners, G. Ziegmann. Developing a new generation of sisal composite fibres for use in industrial applications. Composites Part B: Engineering, 66(2014)287-98.

[10] A. Shalwan, B. Yousif, In state of art: Mechanical and tribological behaviour of polymeric composites based on natural fibres. Mater. Des. 48(2013)14–24.

[11] N. Karthi, K. Kumaresan, S. Sathish, S. Gokulkumar, L. Prabhu, N. Vigneshkumar. An overview: Natural fiber reinforced hybrid composites, chemical treatments and application areas. Materials today: proceedings. 27(2020)2828-34.

[12] M. Abdelmouleh, S. Boufis, M.N. Belgacem, A. Dufresne, Short natural-fibre reinforced polyethylene and natural rubber composites: effect of silane coupling agents and fibre loading. Compos Sci Technol, 67(2007):1627–39.

[13] A.K. Mohanty, M. Misra, L.T. Drzal, Compos Interfaces 8(2001).)313 19.

[14] S. Kumar, A.K. Mohanty, L. Erickson, M. Misra, Lignin and Its Applications with Polymers. J. Biobased Mater. Bioenergy, 3(2009)1–24.

[15] D. Ray, B.K. Sarkar, A.K. Rana, N.R. Bose, Effect of alkali treated jute fibres on composite properties. Bull Mater Sci. 24(2001)129–35.

[16] K.O. Reddy, K.R.N. Reddy, J. Zhang, J. Zhang, A. Varada Rajulu, Effect of alkali treatment on the properties of century fiber. *J Natural Fibers*. *10*(2013)282-296.

[17] S. Kathiresan, O. Meenakshisundaram, Effect of alkali treated and untreated cellulose fibers and human hair on FTIR and tensile properties for composite material applications. *SN Applied Sci. 4*(2022)74.

[18] B.M. Reddy, R.M. Reddy, P.V. Reddy, N.N.A. Prashanth, D. Bandhu, Effect of alkali treatment on mechanical properties and morphology of the Balanites aegyptiaca composite. *Proc Inst Mech Eng, Part C: J Mech Eng Sci.* (2023)09544062231217596.

[19] G. Bernardes, M. de Prá Andrade, M. Poletto, Effect of alkaline treatment on the thermal stability, degradation kinetics, and thermodynamic parameters of pineapple crown fibres. *J Mater Research Technol*, *23*(2023)64-76.

Application of artificial neural network to predict the crystallite size and lattice strain of $NiMnCrMoW_x$ high entropy alloys prepared by powder metallurgy

Naveen Kumar Mindi[1,a*], Jonna Naresh[2,b], Bibhuti Jaipuria[3,c], Syed Nasimul Alam[3,c], Krishna Dutta[3,c]

[1]Department of Mechanical Engineering, Malla Reddy (MR) Deemed to be University, Hyderabad-500100, Telangana, India

[2]Department of Aeronautical Engineering, Tulsiramji Gaikwad-Patil College of Engineering and Technology, Nagpur-441108, Maharashtra, India

[3]Department of Metallurgical and Materials Engineering, National Institute of Technology Rourkela, Rourkela-769008, Odisha, India

[a]mindinaveen331@gmail.com, [b]nareshjonna2020@gmail.com, [c]bibhutijaipuria422@gmail.com, [c]syedn@nitrkl.ac.in, [c]duttak@nitrkl.ac.in

Keywords: High Entropy Alloys, Powder Metallurgy, Crystallite Size, Lattice Strain, Artificial Neural Network

Abstract. In this investigation, $NiMnCrMoW_x$ (x = 0.2, 0.4, 0.6, 0.8,1.0) High Entropy Alloys (HEAs) are synthesized through the powder metallurgy route. To identify the phases and morphology of the milled powders, X-ray diffraction and scanning electron microscope is used. The Crystallite Size (CS) and Lattice Strain (LS) is calculated for the BCC phase obtained in the milled powder alloys by using the Willamson-Hall method and predicted using the Artificial Neural Network (ANN) approach. The CS is reduced from 243 nm to less than 75 nm after 70 h of milling. Whereas the LS increased from 0.45 % to a maximum of 1.91 %. The results from the ANN modeling provided an excellent prediction of CS and LS with outstanding accuracies of 95.27 % and 92.02 %.

1. Introduction

High-entropy alloys have attracted scientific attention in materials science due to their potential mechanical properties, including high strength and hardness, good thermal stability, and excellent resistance to corrosion and wear [1,2]. Compared to conventional alloys, HEAs are formed by mixing five or more principal elements with nearly equiatomic proportions, generally ranging in between 5 and 35 % [3,4]. Although the number of major elements are greater than five, they often develop simple crystal structures like BCC, FCC, or BCC+FCC dual-phase structure [4].

Till date, many of the HEAs are synthesized by the traditional melting methods such as arc melting and induction melting [5]. These techniques are yielding to compositional inhomogeneity and coarse-grained microstructures. In contrast, powder metallurgy technique is capable of producing high-entropy alloys (HEAs) with excellent compositional homogeneity, a broad particle size range, and suitability for mass production [6]. However, HEAs produced by powder metallurgy often show limited mechanical properties due to the wide powder size distribution. Earlier studies have shown that powders milled by high-energy ball mills further refines the microstructure of pre-alloyed powders and to enhance the strength of HEAs through grain boundary strengthening. The main reason behind the strengthening is closely related to the CS of materials. Moreover, CS plays a vital role in the sintering behavior and physical properties of the produced HEAs. A reduced CS helps to enhance sintering ability and possibly lower the required sintering temperature. Therefore, investigating the relationship between CS and milling parameters using both experimental and simulation methods is of significant importance. In recent years, artificial neural network (ANN) modeling increased a lot of interest in materials engineering as an advanced tool for predicting experimental trends without conducting complex experiments. For example, Nagarjuna et al. [7] prepared CoCrFeMnNi HEA by powder metallurgy, and their CS and LS prediction is done using ANN modeling, demonstrating the accuracy of the proposed approach.

In the current study, dual-phase $NiMnCrMoW_x$ (x = 0.2, 0.4, 0.6, 0.8, 1.0 atomic fraction) HEAs is fabricated by powder metallurgy route. The XRD and SEM analysis is carried out to determine the morphological changes and crystallographic properties at different milling times.

Furthermore, prediction of CS and LS of the present study and literature is done by ANN modeling that are prepared by milling at different durations.

2. Experimental Procedure

2.1 Synthesis of HEA Powders

Metallic elemental powders of Mn, Cr, Mo, W with a purity greater than 99 %, and Ni having a purity greater than 99.5 % are procured from Loba Chemie Pvt. Ltd., India. In this investigation, among the five constituent elements, the atomic fraction of W is systematically varied (0.2, 0.4, 0.6, 0.8, and 1.0), while the remaining four elements are kept equiatomic. Based on the preliminary calculations, the metallic powders of required quantity are taken. These powders are milled for a period of 70 h using a high-energy planetary ball mill (Model: Fritsch Pulverisette P-5, Germany). The ball-to-powder ratio is taken as 10:1 with a rotational speed of 300 rpm using tungsten carbide (WC) vials and balls. The balls of diameter 10 mm is used and toluene is used as the process controlling agent (PCA). The milling is automatically stopped for cooling of the milling setup, and powder is taken after every 20 h to observe the phase evolution by using different characterization techniques.

X-ray diffraction (XRD) analysis of the different samples is performed using a Bruker D8 Advance diffractometer equipped with Co Kα radiation (λ = 1.789 Å). The CS and LS are evaluated from the XRD patterns using the Williamson–Hall (W–H) method. The morphology and particle size of the milled powders for different durations are examined by scanning electron microscopy (SEM) to assess the effect of milling time. JEOL JSM-6480 LV microscope operated at an accelerating voltage of 20 kV is used for the powder observation. Elemental composition of the powder samples is determined by energy-dispersive spectroscopy (EDS) integrated with the SEM system.

2.2 Collection of the dataset

The data of CS and LS obtained from this investigation and previously published works are collected to do prediction by using ANN modeling. The literature is chosen from different HEAs synthesized by the planetary ball milling process at different milling times, and then CS and LS are estimated from the XRD results, as shown in Table 1. Out of all the data collected 38 datasets are taken from earlier literature and the remaining 5 are obtained from the present study to know the accuracy of ANN modeling. While collecting data from previously published works, one should keep in mind that the experimental data should be of the same processing technique. To get reliable results the uniformity of experiments is maintained and the data collected from the literature has similar milling conditions. The input parameters such as milling time, ball-to-powder ratio (BPR), and milling speed, are considered. The varied range of CS and LS data corresponds to the milling time, BPR, and milling speed. The compositions chosen for different elements in the HEAs are considered in at. %.

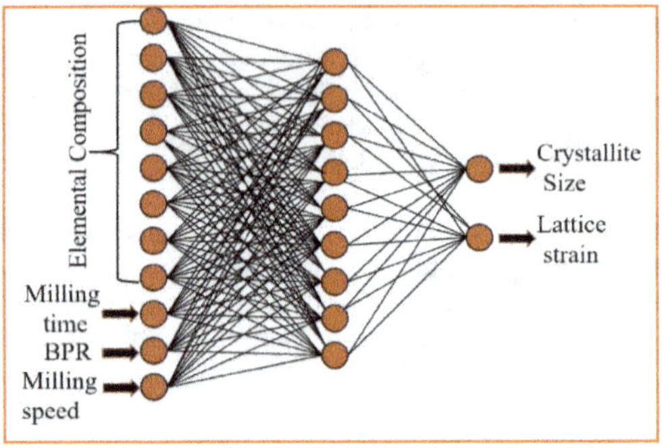

Fig. 1 *The suggested structure of the ANN model for forecasting the CS and LS of HEAs*

Table 1 *Collection of data having elemental composition (at %), milling time (min), milling speed (rpm), and the ball-to-powder ratio.*

S.No	Composition								Parameter			Ref.
	Co	Cr	Fe	Mn	Ni	Al	Mo	W	Time (min)	BPR	Speed	
1	20	20	20	20	20	0	0	0	20	10:1	1100	[8]
2	20	20	20	20	20	0	0	0	60	10:1	1100	[8]
3	16.67	16.67	16.67	16.67	16.67	16.67	0	0	360	15:1	250	[9]
4	16.67	16.67	16.67	16.67	16.67	16.67	0	0	1080	15:1	250	[9]
5	16.67	16.67	16.67	16.67	16.67	16.67	0	0	1800	15:1	250	[9]
6	16.67	16.67	16.67	16.67	16.67	16.67	0	0	2520	15:1	250	[9]
7	20	20	20	20	20	0	0	0	60	10:1	300	[9]
8	20	20	20	20	20	0	0	0	180	10:1	300	[10]
9	20	20	20	20	20	0	0	0	300	10:1	300	[10]
10	20	20	20	20	20	0	0	0	600	10:1	300	[10]
11	20	20	20	20	20	0	0	0	900	10:1	300	[10]
12	20	20	20	20	20	0	0	0	1200	10:1	300	[10]
13	20	20	20	20	20	0	0	0	1800	10:1	300	[10]
14	20	20	20	20	20	0	0	0	2400	10:1	300	[10]
15	20	20	20	20	20	0	0	0	3000	10:1	300	[10]
16	16.67	16.67	16.67	16.67	16.67	16.67	0	0	360	15:1	250	[11]
17	16.67	16.67	16.67	16.67	16.67	16.67	0	0	720	15:1	250	[11]
18	16.67	16.67	16.67	16.67	16.67	16.67	0	0	1080	15:1	250	[11]
19	16.67	16.67	16.67	16.67	16.67	16.67	0	0	1440	15:1	250	[11]
20	16.67	16.67	16.67	16.67	16.67	16.67	0	0	1800	15:1	250	[11]
21	16.67	16.67	16.67	16.67	16.67	16.67	0	0	2520	15:1	250	[11]
22	20	20	20	20	20	0	0	0	1800	15:1	250	[12]
23	20	20	20	20	20	0	0	0	5	10:1	800	[7]
24	20	20	20	20	20	0	0	0	10	10:1	800	[7]
25	20	20	20	20	20	0	0	0	30	10:1	800	[7]
26	20	20	20	20	20	0	0	0	60	10:1	800	[7]
27	20	20	20	20	20	0	0	0	90	10:1	800	[7]

28	20	20	20	20	20	0	0	0	120	10:1	800	[7]
29	20	20	20	20	20	0	0	0	150	10:1	800	[7]
30	20	20	20	20	20	0	0	0	180	10:1	800	[7]
31	20	20	20	20	20	0	0	0	210	10:1	800	[7]
32	20	20	20	20	20	0	0	0	240	10:1	800	[7]
33	20	20	20	20	20	0	0	0	240	10:1	250	[13]
34	20	20	20	20	20	0	0	0	600	10:1	250	[13]
35	20	20	20	20	20	0	0	0	900	10:1	250	[13]
36	20	20	20	20	20	0	0	0	120	10:1	300	[14]
37	20	20	20	20	20	0	0	0	1440	15:1	350	[15]
38	20	20	20	20	20	0	0	0	2100	10:1	200	[16]
39	0	23.8	0	23.8	23.8	0	23.8	4.76	4200	10:1	300	*
40	0	22.72	0	22.72	22.72	0	22.72	9.09	4200	10:1	300	*
41	0	21.73	0	21.73	21.73	0	21.73	13.04	4200	10:1	300	*
42	0	20.83	0	20.83	20.83	0	20.83	16.66	4200	10:1	300	*
43	0	20	0	20	20	0	20	20	4200	10:1	300	*

*Denotes the present study

2.3 Details of modeling

Generally, the ANN model is classified into three different layers: input, hidden, and output. Each layer is associated with a component known as a neuron. To communicate with the neuron the linking weight is used and the change in weights will be carried out by the previous neuron's taking the neuron's input data. The optimization of training, validation, and testing is done using an acceptable backpropagation ANN model depending on several trials for the number of neurons and regulating the number of hidden layers. Further, the model's output data are compared with the original measurements of CS and LS. To determine the absolute value the difference between the predicted and precise values is used. The model has been well-trained and tested to attain extreme potential. After numerous iterations, the model has finally evaluated the outcome. The most appropriate and extensively used algorithm for forecasting the properties of materials is backpropagation which better understands the relationship between input and output datasets. A 11-9-1 structure was used, with 11 neurons in the input layer, 9 neurons in the hidden layer, and 1 neuron in the output layer. The input parameters are the elemental composition of Co, Cr, Fe, Mn, Ni, Al, Mo, and W, while the remaining parameters are milling time, BPR, and milling speed are examined. The elemental composition is given in atomic percentage, and the milling speed is taken in rpm. MATLAB (R2018a) is used for the model's implementation. The suggested backpropagation artificial neural network model for the present study is shown in Fig. 1. The different layers, input, hidden, and output processed by neurons are schematically represented in this diagram.

3. Results and Discussion

3.1 XRD Analysis of Milled HEA Powders

The XRD patterns of $NiMnCrMoW_x$ powders milled for various periods up to 70 h are analyzed. In the starting, all the diffraction peaks of pure elements are seen in the XRD spectrum for all the compositions. With the increase in milling time, there is a substantial decrease in the peak height and broadening of the peak which suggests refinement of powder particle size. For example, considering BCC phase as seen in Fig. 2a, that the molybdenum and tungsten peaks of (110) plane are seen as individual peaks in the initial hours, while after the 70 h of milling it is observed that both the elements merged together. Individual peak position change is observed from Figs. 2 (b and c). This is an indication that after 70 h of milling W and Mo have formed a solid solution resulting in a BCC phase. Similarly, the other BCC elements are alloyed with tungsten element. Moreover, along with BCC phase, a secondary FCC phase is noticed after 70 h of milling which is composed of Ni element formed because of the difference in crystal structure and electronegativity value of nickel-1.91: Pauling scale. The CS and LS is calculated for the BCC phase by using Williamson–Hall method. The Cs and LS are determined from the most intense peak in the XRD spectra by eliminating the instrumental error. The CS is reduced from 243 nm to less than 75 nm after 70 h of milling. Whereas the LS increased from 0.45 % to a maximum of 1.91 %. The detailed explanation can be seen in our earlier publication [17].

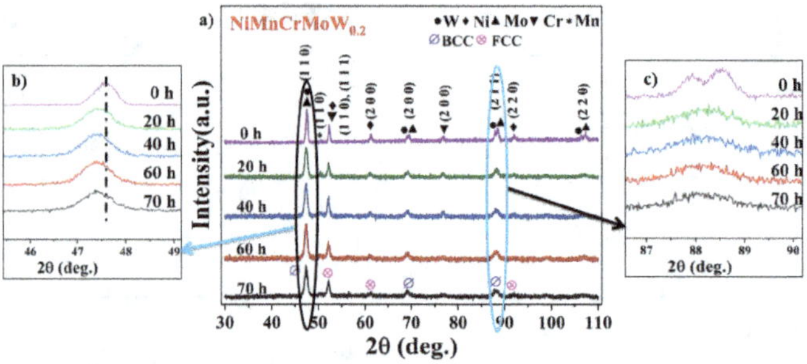

Fig. 2 (a) X-ray diffraction pattern of the milled powders of NiMnCrMoW0.2 for 0, 20, 40, 60, 70h, (b,c) Individual peak position change from 0 h to 70 h milling of (110), (211) planes [17].

3.2 Powder Morphology of HEAs with Milling Time

The analysis of milled powders is investigated from the morphology obtained from SEM as seen from Fig. 3. The SEM image of the unmilled powder mixture from Fig. 3a represents a wide particle size distribution with an average particle size of less than 75 μm. It is noticed from the Fig 3(b-f) that the particles of milled powders are flattened and flake-shaped due to severe plastic deformation. With the increase in W content the percentage of flaky particles is decreased and raise in the round shaped particles. This is mainly because of ductile nature of Mo and Ni lead to flat particle and tungsten which is brittle formed the round particles. In all the compositions, a mixture of both flakes and round shapes is observed and the variation is noticed with W content.

Fig. *3 SEM micrographs of the powder mixture of* $NiMnCrMoW_x$*(x= 0.2, 0.4, 0.6, 0.8, and 1.0) (a) before milling, and after milling(b)* $W_{0.2}$*, (c)* $W_{0.4}$*, (d)*$W_{0.6}$*, (e)* $W_{0.8}$*, (f)* $W_{1.0}$ *[17].*

3.3 Training, Testing, and Validation of the ANN Model

The backpropagation neural network is used to improve the ANN model in this investigation which predicts the CS and LS of the alloys. To maintain consistent input data, all the datasets are made to scale by using the feature scaling function at first. For the ANN model's implementation, the neural network structure chosen was 11-9-1 by using error and trial procedure. The Levenberg Marquardt training algorithm is used for the training of the ANN model. The mean square error (MSE) between predicted and experimental output has been calculated during the development of the ANN model, and the values are equal to 0.00061905 in 24 epochs for CS and 0.001193 in 29 epochs for LS.

3.4 ANN Backpropagation Model to Predict Crystallite Size and Lattice Strain

The ANN backpropagation model is used to create a generalized model to see the combined effect of the Co, Cr, Fe, Mn, Ni, Al, Mo, and W elements in the HEA along with the parameters milling time, speed, and BPR on the CS and LS. After several iterations and training of the ANN model, an exceptional correlation coefficient (R) is calculated for the CS which are R = 0.99577, R = 0.99817, and R = 0.99821 for the training, validation, and testing, respectively. The overall correlation coefficient for the forecast of CS is about R = 0.99588, which is shown in Fig. 4. Whereas, the values of 'R' for training, validation, and testing are obtained as R = 0.99504, R = 0.9916, R = 0.99489, respectively. The overall correlation coefficient for the prediction of LS is achieved as R = 0.994, as presented in Fig. 5. By using the ANN model the absolute error values of CS and LS are predicted as 4.73 % and 7.98 %, respectively showing outstanding agreement between the experimental and predicted data. The current investigation limited the collection of experimental data of CS and LS having similar alloying content and processing route. The precision of the model is appropriate for predicting CS and LS values.

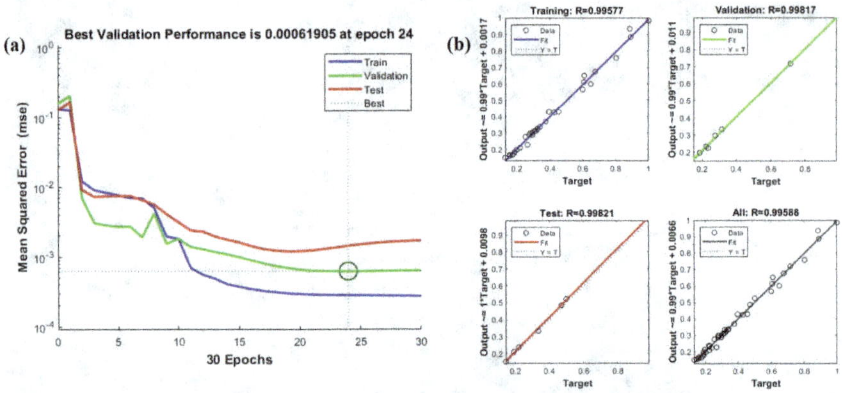

Fig. 4 (a) Mean square error value of CS depending on the best validation performance, (b) Correlation coefficient of CS during the modeling.

The CS and LS values of Experimental and Predicted are shown in Fig. 6. It is significant to know that the created ANN model predicts the desired CS and LS of the present HEAs without conducting any experiments. ANN model used for the prediction of CS and LS provides exceptional accuracy with the experimental results achieved from this investigation and other HEAs reported [8-16]. As we know the CS plays a crucial role in the sintering and its physical properties. Therefore, the prediction of CS of particular HEAs using the defined milling conditions provides an idea so that we can minimize the experimental time and cost during synthesis of HEAs.

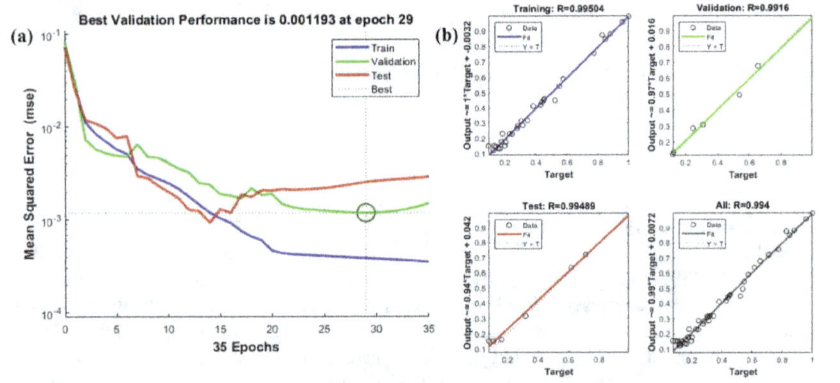

Fig. 5 (a) Mean square error value of LS depending on the best validation performance, (b) Correlation coefficient of LS during the modeling.

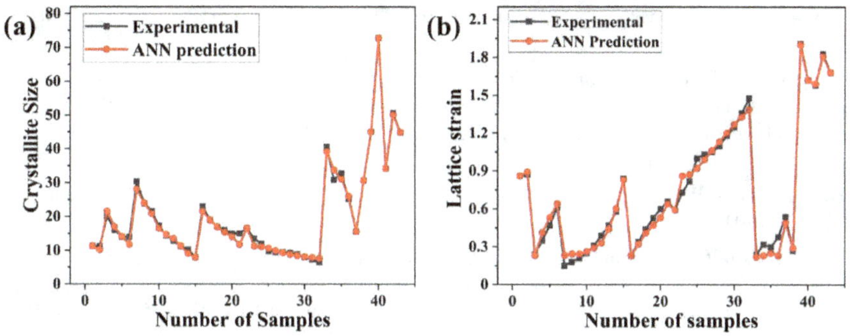

Fig. 6 Representation of experimental and predicted values using the ANN approach of (a) Crystallite Size, and (b) Lattice strain.

Conclusion

In the current study, $NiMnCrMoW_x$ HEA powders are developed using through the powder metallurgy route by using mechanical alloying technique. From the XRD results it is observed that a dual-phase (a mixture of BCC and FCC) structure is formed after milling upto 70 h. The CS and LS is determined for the BCC phase occurred for the 70 h milled powders by using the Willamson-Hall method. Also, the prediction of CS and LS is done by using the Artificial Neural Network (ANN) approach. The SEM images shown the transformation of flake particle to round with increase in tungsten content. After 70 h of milling, the CS is reduced from 243 nm to less than 75 nm. While the LS increased to 1.91 %. The ANN modeling provided a prediction of CS and LS with an accuracy of 95.27 % and 92.02 % respectively.

Credit authorship contribution statement

Naveen Kumar Mindi: Writing – original draft, Experimentation, and data curation, Conceptualization, Investigation, Methodology, **Syed Nasimul Alam:** Supervision, Conceptualization, Writing - review & editing, **Krishna Dutta:** Resources, Supervision, Conceptualization, Writing - review & editing, **Jonna Naresh:** Review & editing, **Bibhuti Jaipuria:** Review & editing.

Declaration of competing interest

Naveen Kumar Mindi reports administrative support and equipment, and supplies were provided by the National Institute of Technology Rourkela. Further, the authors declare that they have no known competing financial interests or personal relationships that could have appeared to influence the work reported in this paper.

Acknowledgments

The authors would like to acknowledge the DST-FIST-supported XRD laboratory in the Dept. of Metallurgical and Materials Engg, at NIT Rourkela for their support in the phase analysis of the developed high entropy alloys.

References

[1] Y. Zhang, T. T. Zuo, Z. Tang, M. C. Gao, K. A. Dahmen, P. K. Liaw, Z. P. Lu, Microstructures and properties of high-entropy alloys, Prog. Mater. Sci. 61 (2014) 1-93. https://doi.org/10.1016/j.pmatsci.2013.10.001.

[2] A. Sharma, H. Lee, B. Ahn, Effect of Additive Elements (x = Cr, Mn, Zn, Sn) on the Phase Evolution and Thermodynamic Complexity of AlCuSiFe-x High Entropy Alloys Fabricated via

Powder MetallurgyMet. Mater. Int. 28 (2022) 2216-2224. https://doi.org/10.1007/S12540-021-01125-0.

[3] J. W. Yeh, S. K. Chen, S. J. Lin, J. Y. Gan, T. S. Chin, T. T. Shun, C. H. Tsau, and S. Y. Chang, Nanostructured high-entropy alloys with multiple principal elements: novel alloy design concepts and outcomes, Adv. Eng. Mater. 06(5) (2004) 299–303. https://doi.org/10.1002/adem.200300567

[4] B. S. Murty, J. W. Yeh, S. Ranganathan, High-entropy alloys, Butterworth-Heinemann, London, UK (2014).

[5] Y. A. Alshataif, S. Sivasankaran, F. A. Al-Mufadi, A. S. Alaboodi, H. R. Ammar, Manufacturing Methods, Microstructural and Mechanical Properties Evolutions of High-Entropy Alloys: A Review, Met. Mater. Int. 26 (2020) 1099. https://doi.org/10.1007/s12540-019-00565-z

[6] C. Suryanarayana, Mechanical alloying and milling, Prog. Mater. Sci. 46 (2001) 1-184. https://doi.org/10.1016/S0079-6425(99)00010-9

[7] Ch. Nagarjuna, S. K. Dewangan, A. Sharma, K. Lee, S. J. Hong, B. Ahn, Application of Artifcial Neural Network to Predict the Crystallite Size and Lattice Strain of CoCrFeMnNi High Entropy Alloy Prepared by Powder Metallurgy, Met. Mater. Int. 29 (2023) 1968–1975. https://doi.org/10.1007/s12540-022-01355-w

[8] S.H. Joo, H. Kato, M.J. Jang, J. Moon, E.B. Kim, S.J. Hong, and H.S. Kim, Structure and properties of ultrafine-grained CoCrFeMnNi highentropy alloys produced by mechanical alloying and spark plasma sintering, J. Alloys Compd. 698 (2017) 591-604. http://dx.doi.org/10.1016/j.jallcom.2016.12.010

[9] W. Ji, Z. Fu, W. Wang, H. Wang, J. Zhang, Y. Wang, F. Zhang, J. Alloys Compd. 589 (2014) 61.

[10] X. Liu, H. Yin, and Y. Xu, Microstructure, Mechanical and tribological properties of oxide dispersion strengthened high-entropy alloys, J. Mat. 10 (2017) 1312, http://dx.doi.org/10.3390/ma10111312

[11] C. Wang, W. Ji, and Z. Fu, Mechanical alloying and spark plasma sintering of CoCrFeNiMnAl high-entropy alloy, Adv. Powder Technol. 25 (2014) 1334–1338. http://dx.doi.org/10.1016/j.apt.2014.03.014

[12] W. Ji, W. Wang, H. Wang, J. Zhang, Y. Wang, F. Zhang, and Z. Fu, Alloying behavior and novel properties of CoCrFeNiMn high-entropy alloy fabricated by mechanical alloying and spark plasma sintering, Intermetallics. 56 (2015) 24-27. http://dx.doi.org/10.1016/j.intermet.2014.08.008

[13] Y. Liu, J. Wang, Q. Fang, B. Liu, Y. Wu, and S. Chen, Preparation of superfine-grained high entropy alloy by spark plasma sintering gas atomized powder, Intermetallics. 68 (2016) 16-22. http://dx.doi.org/10.1016/j.intermet.2015.08.012.

[14] D. Yim, P. Sathiyamoorthi, S.J. Hong, and H.S. Kim, Fabrication and mechanical properties of TiC reinforced CoCrFeMnNi high-entropy alloy composite by water atomization and spark plasma sintering, J. Alloys Compd. 781 (2019) 389-396. https://doi.org/10.1016/j.jallcom.2018.12.119

[15] H. Hadraba, Z. Chlup, A. Dlouhy, F. Dobes, P. Roupcova, M. Vilemova, and J. Matejicek, Oxide dispersion strengthened CoCrFeNiMn high-entropy alloy, Mater. Sci. Eng. A 689 (2017) 252–256, http://dx.doi.org/10.1016/j.msea.2017.02.068

[16] L. Rogal, D. Kalita, A. Tarasek, P. Bobrowski, and F. Czerwinski, Effect of SiC nano-particles on microstructure and mechanical properties of the CoCrFeMnNi high entropy alloy, J. Alloys Compd. 708 (2017) 344-352. http://dx.doi.org/10.1016/j.jallcom.2017.02.274

[17] N.K. Mindi, S.N. Alam, K. Dutta, Effect of Tungsten on Microstructure and Mechanical Properties of Novel NiMnCrMoWx High-Entropy Alloys Developed by Mechanical Alloying, J. Mater. Eng. Perform. 34 (2025) 13634-13651. https://doi.org/10.1007/s11665-024-10107-3

Grain-scale physics to array performance: HZO-Driven FeFET modelling for neuromorphic computing

Aditya Ram ABBARAJU[1,a], Anandhan SRINIVASAN[1,b], Balasubramanian KANDASUBRAMANIAN[2,c*]

[1]Department of Metallurgical and Materials Engineering, NIT Karnataka, Surathkal- 575025, Karnataka, India

[2]Department of Metallurgical and Materials Engineering, Defence Institute of Advanced Technology (DU), Girinagar, Pune- 411025, Maharashtra, India

[a]aadityaram.242nt001@nitk.edu.in, [b]anandhan@nitk.edu.in, [c]meetkbs@gmail.com

Keywords: In-Memory Computing, FeFET Crossbar, Microstructure Variability, HZO Ferroelectric, Phase-Field Modelling, Physics-Informed ML

Abstract. Ferroelectric $Hf_{0.5}Zr_{0.5}O_2$ (HZO) exhibits grain-size-dependent polarization switching that dominates variability in FeFET neuromorphic crossbars, yet the quantitative link from microstructure to array-level accuracy remains unexplored. This work bridges polycrystalline grain physics to 64×64 array performance via our proposed grain count conditioned surrogate, *GrainAwareNet,* developed using a 2D cross-sectional phase-field model assuming columnar grain structure typical of thin HZO films. Simulations yield *1,927 valid P-E loops* spanning grain counts G = 10-100 (equivalent mean grain size ~6-19 nm), with P_r = 15-35 μC/cm² (±15%) and E_c = 80-180 MV/m (±40%). GrainAwareNet achieves $R^2 = 0.9923$, RMSE = 0.032 μC/cm², and *1000× speedup*, enabling Monte Carlo exploration of grain-count variability σ_G = 0-40. The framework demonstrates *remarkable robustness*: Matrix-Vector Multiply (MVM) output error remains *<1% for σ_G up to ~35*, with *<0.3% error at σ_G = 5* (CV < 6%). Voltage calibration further recovers ~90% accuracy at higher variability. Framework enables microstructure-aware design-technology co-optimization (DTCO) for HZO FeFETs.

1. Introduction

Brain-inspired neuromorphic computing promises orders-of-magnitude energy efficiency gains over conventional digital systems [1]. Ferroelectric field-effect transistors (FeFETs) offer compelling hardware synapses with tunable, nonvolatile conductance, enabling on-chip learning via voltage-gated polarization switching [2]. HZO-based ferroelectrics enable CMOS-compatible integration (annealing ≤600 °C, TiN/W electrodes, 10 nm scalability), unlike perovskite predecessors requiring noble metals and thick films [3]. However, polycrystalline HZO contains heterogeneous grains whose boundaries suppress ferroelectricity, introducing device-to-device variability that propagates to crossbar compute precision [4,5].

Prior compact and phase-field modeling efforts have established physically grounded ferroelectric switching [1,5], yet they suffer from critical limitations in design-technology co-optimization (DTCO). Existing flows typically assume abstract Gaussian noise to model variability, ignoring its physical origin in microstructure [3]. This abstraction prevents the predictive optimization of process control targets, such as grain size, needed to meet specific accuracy thresholds. Furthermore, direct physical simulations remain too computationally expensive to bridge the gap between grain-scale physics and large-scale array metrics. This paper closes the gap by combining (i) grain-resolved TDGL simulations to generate a statistically robust dataset, (ii) a novel surrogate model conditioned on *scalar grain statistics* (G) to accelerate sampling, and (iii) crossbar-level Monte Carlo analysis to directly quantify MVM output fidelity under realistic grain distributions.

2. Methodology

2.1 Grain-Scale Physics and Dataset Generation

2.1.1 Polycrystalline HZO Microstructure Model

Model Scope and Assumptions: This work employs a 2D cross-sectional phase-field model assuming a *columnar grain microstructure*, where individual grains extend uniformly through the ~10 nm film thickness (z-axis). This assumption is well-validated for thin HZO films deposited via ALD/PVD, where transmission electron microscopy (TEM) studies confirm that grains nucleate at the bottom interface and grow vertically to span the full thickness [4,6]. Under this geometry, in-plane grain-boundary effects (modeled in 2D) dominate polarization variability, while through-thickness domain evolution remains approximately uniform. The 2D approach is computationally tractable while capturing the primary microstructure-driven variability observed experimentally [4].

Polycrystalline HZO is modeled with a time-dependent Ginzburg-Landau (TDGL) evolution of polarization order parameter $P(r, t)$ under external field E, where grain interiors favor ferroelectric ordering ($\alpha_{ferro} < 0$) while grain boundaries suppress ferroelectricity via altered local energetics ($\alpha_{para} > 0$, "boundary pinning") [5]. The discretized evolution on a 60 nm × 60 nm domain (N = 60, Δx = 1 nm, periodic boundaries) follows (Eq. 1).

$$\partial P/\partial t = -\rho\, [\, \alpha(G)\, P + \beta\, P^3 + \gamma\, P^5 - g_{11}\, \nabla^2 P - E_{ext}\,] \tag{1}$$

The TDGL equation is solved in normalized units where polarization P is in μC/cm² and electric field E is in MV/m. The Landau coefficients are *effective phenomenological parameters* calibrated to reproduce the experimental coercive field (E_c ~ 1-2 MV/cm) and saturation polarization (P_s ~ 25 μC/cm²) of HZO films [4,5]Specifically, $\alpha(G)$ applies a *mean-field scaling* to grain interiors, approximating the macroscopic Hall-Petch-like hardening observed in fine-grained films [6], while α_{para} explicitly models local domain wall pinning at boundaries. This separates global size effects from local topology effects.

2.1.2 Dataset Generation and Physical Validation

A Monte Carlo workflow generates 2,000 target simulations across grain counts G = 10-100 (uniform random), retaining *1,927 valid P-E loops* after numerical screening. Each simulation applies cyclic triangular field sweeps E(t) with peak amplitude E_{max} drawn from $\mathcal{N}(5.0\times10^8$ V/m, 10% noise) over 4 ns total. Post-processing extracts P_r (remanent polarization) and E_c (coercive field), confirming physical bounds ($P_{sat} \approx 30$ μC/cm², $E_c \in$ [80-180] MV/m).

The dataset spans *P_r = 15-35 μC/cm² (±15%) and E_c = 80-180 MV/m (±40%)*, capturing grain-conditioned hysteresis diversity (Fig. 1). Soft grains ($G < 20$, mean diameter $D \approx 13.4$ nm) show higher P_r and lower E_c; hard grains ($G > 80$, $D \approx 6.7$ nm) show lower P_r and higher E_c, validating the grain-physics coupling [4].

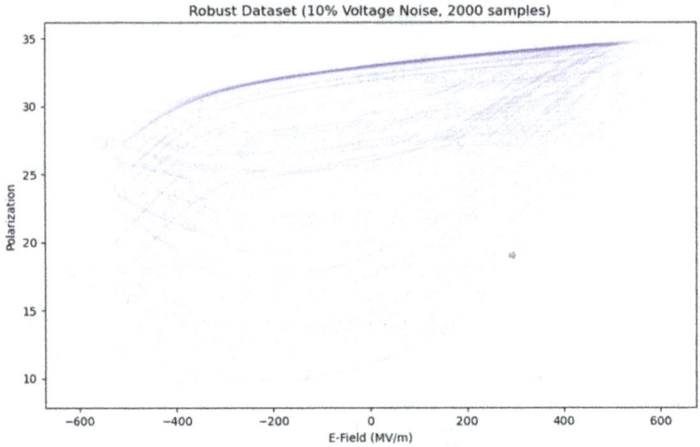

Fig. 1. Generated dataset of 1,927 valid P-E hysteresis loops showing variability in Remanent Polarization (P_r) and Coercive Field (E_c) driven by grain microstructure (G = 10-100).

Table 1. Comparison of perovskite and hafnia-based ferroelectric material comparison.

Property	Perovskites (PZT, $BaTiO_3$)	Hafnia-Based HZO
Annealing Temperature	~700 °C	400-600 °C
Electrodes	Pt, $SrRuO_3$	TiN, W (CMOS)
Thickness Scalability	~100 nm	~10 nm
Coercive Field Ec	0.05-0.25 MV/cm	0.8-2 MV/cm
Endurance	~10^{10} cycles	≥10^{10} cycles

2.2 GrainAwareNet: Physics-Informed Surrogate

2.2.1 Architecture and Grain Injection

Direct TDGL rollouts require ~3 minutes per loop; this study proposes *GrainAwareNet*, trained to predict incremental polarization updates ΔP conditioned on instantaneous state (P, E, ΔE) and the *scalar microstructure statistic G (grain count). Note that conditioning on the scalar count effectively predicts the ensemble-averaged polarization path for a given grain size, smoothing out specific topological variations*. The key contribution is *grain injection at every hidden layer*, preserving microstructure sensitivity. The network uses three fully connected layers (64→128→64 neurons) with tanh activation and grain concatenation.

2.2.2 Training and Validation

Trained on 80% of the dataset (1,542 loops), GrainAwareNet achieves $R^2 = 0.9923$ *and RMSE* = *0.032 μC/cm²* on held-out data. A sensitivity diagnostic yields 6.8×10^{-3} *μC/cm² per grain unit* (Fig. 2), confirming grain-awareness. Validation is against the TDGL model, calibrated to experimental literature [4,7]. Although the surrogate is trained on nanosecond-scale loops, we utilize the *quasi-static assumption* for array operations. Since the write pulse (~3 *μs)* is significantly slower than intrinsic domain dynamics, the surrogate accurately predicts the final equilibrium polarization state regardless of pulse duration.

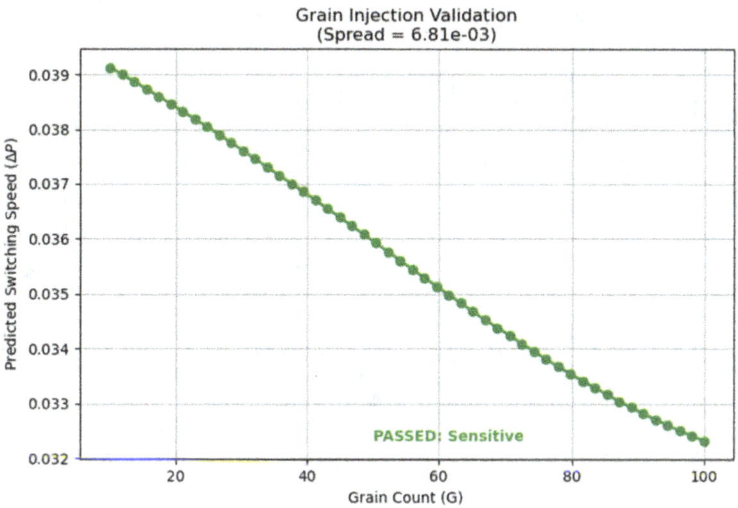

Fig. 2. Grain sensitivity diagnostic: ΔP versus G at fixed field. Spread = 6.8×10^{-3} $\mu C/cm^2$ confirms grain injection success.

Computational speedup: ~0.2 ms/loop vs ~3 min/loop enables *~1000× speedup* for Monte Carlo exploration.

2.3 Crossbar Array Simulation and Critical Threshold

2.3.1 Write Protocol and Monte Carlo Setup

We simulated a *64×64 FeFET crossbar* with differential-pair encoding ($w = g_+ - g_-$). For each device, grain count is drawn from (50, σ_G) and clipped to [10, 100]. A write pulse $E_{max} = 3.5\times10^8$ V/m is applied in 325 steps (~3 µs), with surrogate rollout predicting final polarization. For readout, we assume a *linear mapping* between the programmed ferroelectric polarization and the channel conductance. *MVM output error* is computed as given in (Eq. 2).

$$\text{MVM Error} = 100 \times ||\mathbf{Y}_{ideal} - \mathbf{Y}_{real}||^2F \,/\, ||\mathbf{Y}_{ideal}||^2F \tag{2}$$

where $Y = WX$. Five Monte Carlo runs per σ_G record mean and standard deviation.

2.3.2 Robustness Analysis

The array demonstrates *graceful degradation* (Fig. 3): MVM error remains *<0.3% at $\sigma_G = 5$ (CV < 6%), rises to ~0.4% at $\sigma_G = 20$, and crosses 1% only beyond $\sigma_G \approx 35$.* This robustness arises from: Differential pair encoding (partial mismatch cancellation) - Statistical averaging (64×64 array reduces single-device impact) - Surrogate smoothing ($R^2 = 0.9923$ averages grain extremes) For context, experimental HZO processes exhibit CV ≈ 10-15% ($\sigma_G \approx$ 15-20) [4,6], well within the <1% error regime.

Materials Research Proceedings 65 (2026) 74-79 https://doi.org/10.21741/9781644904138-11

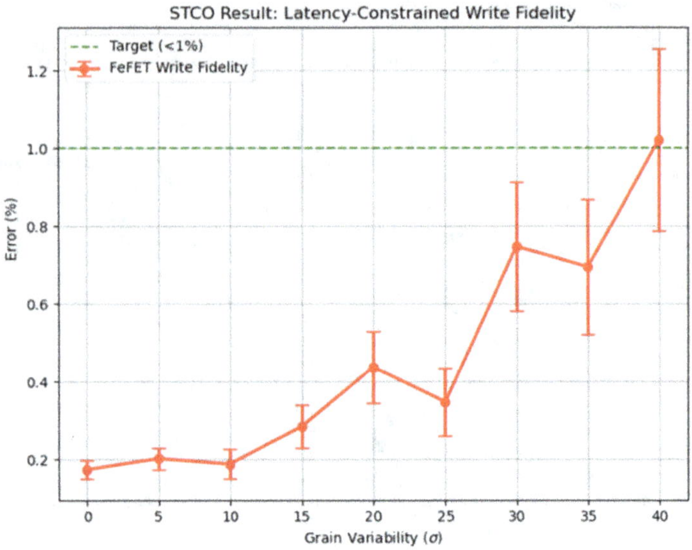

Fig. 3. MVM output error versus σ_G. Error remains <1% up to $\sigma_G \approx 35$, demonstrating robustness to grain variability.

3. Results and Discussion

3.1 Surrogate Accuracy and Computational Benefit

Table 2. GrainAwareNet performance and array robustness metrics.

Metric	**Value**
Phase 1: Valid loops	1,927
Phase 1: Grain range *G* (equiv. *D* nm)	10-100 (6-19 nm)
Phase 2: Test R^2	0.9923
Phase 2: Grain sensitivity	6.8×10^{-3} μC/cm²
Phase 2: Speedup vs TDGL	~1000×
Phase 3: MVM error at σ_G=5	*<0.3%*
Phase 3: MVM error at σ_G=20	*~0.4%*
Phase 3: 1% error threshold	$\sigma_G \approx 35$

The *1000× speedup* unlocks design-space exploration infeasible with physics simulators alone.

3.2 Voltage Calibration

Adaptive voltage calibration (±10% *E*max adjustment) yields *~90% accuracy recovery* at $\sigma_G = 20$, demonstrating circuit-level mitigation can further relax process requirements.

Conclusion

This study presents the first integrated framework connecting polycrystalline grain microstructure to neuromorphic array compute precision via *GrainAwareNet* ($R^2 = 0.9923$, 1000× speedup). The results quantify *robust operation*: MVM output error remains *<1% for σ_G up to ~35*(CV ≈ 20%), encompassing typical experimental HZO process distributions. For applications demanding near-ideal precision (<0.5% error), $\sigma_G < 20$ is recommended. This framework bridges materials physics to system-level DTCO, enabling predictive design of HZO-based neuromorphic hardware. It is noted that by conditioning on the scalar grain count, the surrogate effectively predicts the *ensemble-averaged* response for that grain size, filtering out intra-device topological noise. Thus, the reported variability metrics represent a *lower bound*, capturing the dominant grain-size-driven mismatch. Future work could extend to wake-up/fatigue modeling and 3D grain-growth kinetics.

References

[1] R. Koduru, I. Ahmed, A.K. Saha, X. Lyu, P. Ye, S.K. Gupta, Phase-field simulations of polarization variations in polycrystalline Hf0.5Zr0.5O2 based MFIM: Voltage dependence and dynamics, J. Appl. Phys. 134 (2023) 084104. https://doi.org/10.1063/5.0158997

[2] A.-L. Kévin, D. Damien, G. Brice, Ultrafast and accurate prediction of polycrystalline hafnium oxide phase-field ferroelectric hysteresis using graph neural networks, Nanoscale Adv. 6 (2024) 2350–2362. https://doi.org/10.1039/D3NA01115A

[3] C. Wang, J. Victor, A.K. Saha, X. Chen, M. Si, T. Sharma, K. Roy, P.D. Ye, S.K. Gupta, FeFET-Based Synaptic Cross-Bar Arrays for Deep Neural Networks: Impact of Ferroelectric Thickness on Device-Circuit Non-Idealities and System Accuracy, in: 2023 Device Res. Conf. DRC, IEEE, Santa Barbara, CA, USA, 2023: pp. 1–2. https://doi.org/10.1109/DRC58590.2023.10187042

[4] H. Chen, H. Luo, X. Yuan, D. Zhang, Constructing a correlation between ferroelectricity and grain sizes in Hf0.5 Zr0.5 O2 ferroelectric thin films, CrystEngComm 24 (2022) 1731–1737. https://doi.org/10.1039/D1CE01626A

[5] A.K. Saha, K. Ni, S. Dutta, S. Datta, S. Gupta, Phase field modeling of domain dynamics and polarization accumulation in ferroelectric HZO, Appl. Phys. Lett. 114 (2019) 202903. https://doi.org/10.1063/1.5092707

[6] H.J. Kim, M.H. Park, Y.J. Kim, Y.H. Lee, W. Jeon, T. Gwon, T. Moon, K.D. Kim, C.S. Hwang, Grain size engineering for ferroelectric Hf0.5Zr0.5O2 films by an insertion of Al2O3 interlayer, Appl. Phys. Lett. 105 (2014) 192903. https://doi.org/10.1063/1.4902072

[7] S. Sugathan, K. Thekkepat, S. Bandyopadhyay, J. Kim, P.-R. Cha, A phase field model combined with a genetic algorithm for polycrystalline hafnium zirconium oxide ferroelectrics, Nanoscale 14 (2022) 14997–15009. https://doi.org/10.1039/D2NR02678C

Heterogeneous development of twinning characteristics in titanium subjected to directional bending

Sujit Nayak[1a*], Somjeet Biswas[1b]

[1]Light Metals and Alloys Research Lab, Department of Metallurgical and Materials Engineering, Indian Institute of Technology, Kharagpur, West Bengal, 721302, India

[a]sujitnayak16@gmail.com, [b]somjeetbiswas@gmail.com

Keywords: Titanium, Bending, Anisotropy, Twinning

Abstract. The present study investigates the through-thickness heterogeneous twin evolution in a hot-rolled titanium (Ti) sheet subjected to 90° bending along the rolling (RD) and transverse (TD) direction at room temperature. The microstructure evolution of the bent sheets was investigated using a field-emission scanning electron microscope (FESEM) equipped with electron backscatter diffraction (EBSD). Significant variations in twin formation were observed in both the bend loading directions and in through-thickness regions. The TD bend sample showed a dominant $\{10\bar{1}2\}\langle 10\bar{1}1\rangle$ extension twins (ET) in the inner compressive region (ICR), with few $\{11\bar{2}2\}\langle 11\bar{2}\bar{3}\rangle$ contraction twins (CT) in the outer tensile region (OTR). In contrast, the RD bend sample exhibited $\{10\bar{1}2\}$ ET and $\{11\bar{2}2\}$ CT in both the regions, but with a lower overall twin fraction than in the ICR of the TD bend sample. The present findings highlight the critical roles of bend-loading direction and regional stress states in governing heterogeneous twin evolution.

Introduction

Titanium (Ti) and its alloys are considered highly suitable as structural components for space, aircraft, marine, defence, and biomedical applications. This is due to its high specific strength, formability, excellent corrosion resistance and biocompatibility [1]. Most of the structural components are made from sheets that require numerous forming operations, often involving bending. During the bending operation, the sample undergoes tensile and compressive loading in the outer and inner regions, respectively, resulting in a through-thickness distribution of stress and strain [2,3]. Owing to the hexagonal close-packed (HCP) crystal structure, α-Ti exhibits high anisotropic in deformation behavior [4]. Due to the low c/a ratio, plastic deformation in α-Ti predominantly occurs through prismatic $\langle a\rangle$ slip, followed by basal $\langle a\rangle$ and pyramidal $\langle a\rangle$ at room temperature (RT) [5]. Along with this, twinning must be initiated to accommodate the strain along the c-axis [5]. At RT, $\{10\bar{1}2\}\langle 10\bar{1}1\rangle$ extension twin (ET) and $\{11\bar{2}2\}\langle 11\bar{2}\bar{3}\rangle$ contraction twin (CT) primarily activates under c-axis tension and compression, respectively [4,6]. The activation of ET and CT led to a lattice rotation by $\sim 85^\circ$ about $\langle 11\bar{2}0\rangle$ and $\sim 63.5^\circ$ about $\langle 10\bar{1}0\rangle$, w.r.t. parent matrix, respectively. This lattice rotation may change the Taylor factor and affect the slip activities in the twin domain compared to the parent matrix under the same external loading [7]. For instance, during the compression of extruded α-Ti in the axial direction, the increase in the strain hardening rate in stage II could be attributed to the activation of $\{10\bar{1}2\}\langle 10\bar{1}1\rangle$ ET twins [6]. However, the activation of $\{11\bar{2}2\}\langle 11\bar{2}\bar{3}\rangle$ CT leads to geometric softening and a decrease in the strain hardening rate [7]. Considering the texture-dependent twinning behavior stated above, the bending behavior of titanium, characterized by a pronounced stress-strain gradient, needs to be explored.

Aiming at the above, the present work explores the three-point bending behavior of α-Ti at RT. The microstructural features were investigated at different regions of the bending sample using a field-emission scanning electron microscope (FESEM) equipped with an electron back-scattered diffraction (EBSD) (Zeiss Sigma 300 VP). The present investigation correlates the heterogeneous microstructure evolution across the thickness with the activation of different deformation twins, as governed by the initial texture during bending.

Materials and Methods

The material used in the present study was in the form of hot-rolled grade II α-Ti sheets. Samples with a thickness of $2\ mm$, width of $20\ mm$ and a length of $50\ mm$ were cut along the rolling direction (RD) and the transverse direction (TD) and referred to as the TD and RD bend samples, respectively. The three-point bend tests were carried out up to a bend angle of ~90° with a strain rate of $0.005\ S^{-1}$ at RT. After the bend test, the samples were sectioned at the mid-width direction to reveal the ND-RD and ND-TD planes for the TD and RD bend samples, respectively. The microstructural features were characterized using the SEM-EBSD technique. For this, the samples were prepared by standard metallographic polishing using emery papers, followed by colloidal silica ~0.05μm. After that, the samples were electropolished with an A3 solution for 15 seconds at 18V. The obtained results were processed using TSL OIM 8 software.

Results and discussions

Microstructure of the initial material

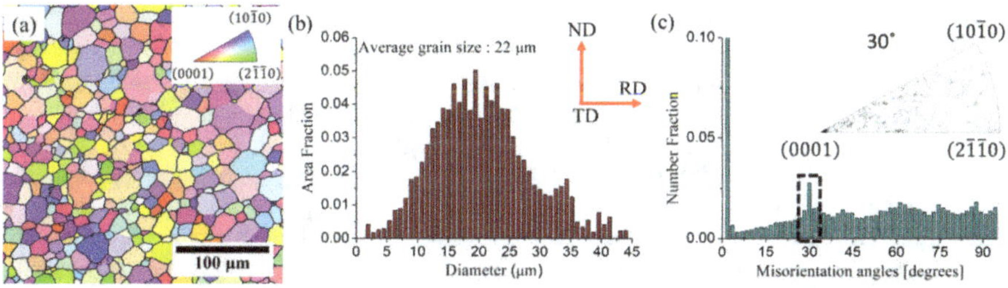

Fig. 1. Microstructure characteristics of the initial hot-rolled α-Ti on ND-RD plane in terms of (a) Inverse pole figure (IPF) map, (b) grain size distribution, and (c) misorientation angle distribution.

Fig. 1a represents the microstructural characteristics of the initial hot-rolled α-Ti in terms of the inverse pole figure (IPF) map obtained on the ND-RD plane. The colors of the IPF map represent the orientation of the grains with respect to the color key code mentioned in the inset of Fig. 1a. The initial material had an equiaxed grain structure with no twins, Fig. 1a. The initial material had a log-normal grain size distribution with a range of $1 - 50$ μm as represented in Fig. 1b, with an area fraction average grain size of ~22 μm. Fig. 1c represents the misorientation angle distribution (MAD) plotted over a range of $1 - 95°$. The plot shows a local maximum at ~30°, which has an orientation relationship with the $[0002]$ axis due to the presence of recrystallized grains [8].

Microstructure evolution in the TD bend sample

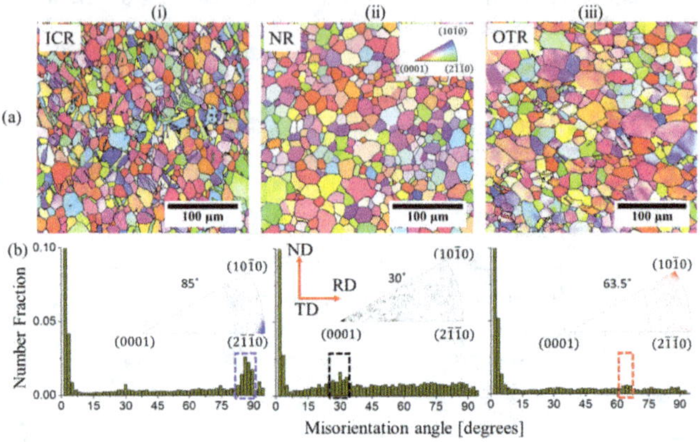

Fig. 2. Microstructure characteristics of different regions for the TD bend sample in terms of IPF maps and misorientation angle distribution of i(a, b) ICR, ii(a, b) NR, and iii(a, b) OTR.

Fig. 2 shows the microstructural characteristics of the TD bend sample using the IPF map, MAD plot of different regions on the ND-RD plane. The bend regions could be distinguished as the inner compression region (ICR), the neutral region (NR), and the outer tension region (OTR). A large fraction of twins could be observed in the ICR of the TD sample (Fig. 2a(i)). The corresponding MAD plot in Fig. 2b(i) shows a local maximum corresponding to 85° peak, which has an orientation relationship with $\langle 11\bar{2}0 \rangle$ axis as illustrated by the angular IPF plot in the inset of Fig. 2b(i). This indicates that the observed twins correspond to $(10\bar{1}2)\langle 10\bar{1}1 \rangle$ ET. However, no significant changes could be observed for the NR (Fig. 2a(ii)), and the microstructure remains equiaxed, similar to that of the initial material (Fig. 1a). The MAD plot in Fig. 2b(ii) also confirms no sign of twin formation; however, the 30° peak shows an orientation relationship with respect to [0002] axis due to the presence of recrystallized grains [8]. In contrast, some of the equiaxed grains of the initial material were elongated for the OTR, and their elongated directions were parallel to the tensile direction (Fig. 2a(iii)). Additionally, a few twins could also be observed. The MAD plot (Fig. 2b(iii)) shows a local maximum at ~63.5° which has an orientation relationship with $\langle 10\bar{1}0 \rangle$ axis. This indicates that the observed twins were $\{11\bar{2}2\}\langle 11\bar{2}\bar{3} \rangle$ CTs.

Microstructure evolution in the RD bend sample

Fig. 3 shows the microstructure characteristics of the RD bend sample using the IPF map and MAD plot of different regions on the ND-TD plane. A large fraction of twins could be observed in the ICR of the RD bend sample (Fig. 3a(i)). The MAD plot in Fig. 3b(i) shows a local maximum at ~63.5°, which is aligned with the $\langle 10\bar{1}0 \rangle$ axis, and at ~85°, which has an orientation relationship with $\langle 11\bar{2}0 \rangle$ axis. This indicates the observed twins correspond to $\{11\bar{2}2\}\langle 11\bar{2}\bar{3} \rangle$ CTs and $(10\bar{1}2)\langle 10\bar{1}1 \rangle$ ETs, respectively. However, the fraction of ETs formation was found to be higher than that of CTs (Fig. 3b(i)). Therefore, both ET and CT were operative in the ICR of the RD bend sample, but a suppression of ET activity could be observed compared to the ICR of the TD bend sample (Fig. 2b(i)). Fig. 3a(ii) represents the microstructure of the NR for the RD bend sample. The map shows an equiaxed grain structure with no twins, suggesting that the NR underwent no substantial deformation and is similar to that of the initial material. Further, this could be confirmed by the MAD plot in Fig. 3b(ii).

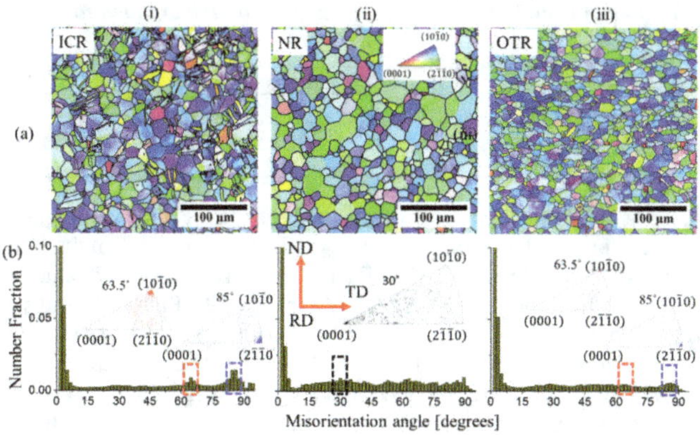

Fig. 3. Microstructure characteristics of different zones for the TD bend sample in terms of IPF maps and misorientation angle distribution of i(a, b) ICR, ii(a, b) NR, and iii(a, b) OTR.

In contrast, the grains in the OTR of the RD bend sample were elongated (Fig. 3a(iii)), similar to those of the OTR of the TD bend sample (Fig. 2a(iii)). Additionally, a few twins could also be observed. The MAD plot reveals that the observed twins have a misorientation angle/axis pair of $63.5°\langle 10\bar{1}0\rangle$ and $85°\langle 11\bar{2}0\rangle$, which corresponds to the CT and ET, respectively (Fig. 3b(iii)). However, the fraction of ET and CT in the OTR was lower than the ICR of the RD bend sample.

Effect of initial texture

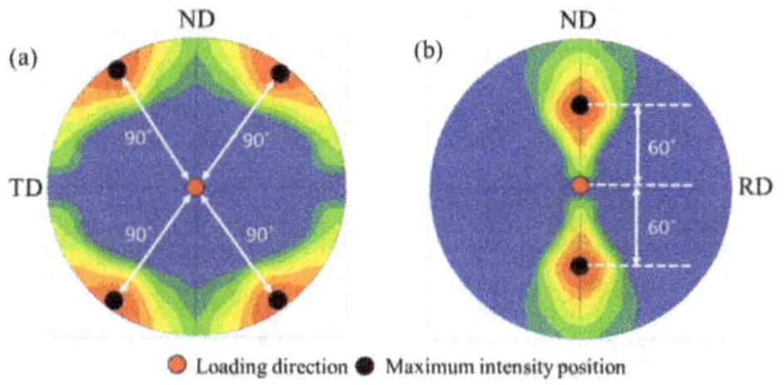

Fig. 4. (0002) pole figures obtained from the cross-sectional area of the initial undeformed (a) TD, (b) RD samples.

The heterogeneous twinning observed in the different regions of the TD and RD bend sample can be understood in terms of the grain orientation (texture) and the nature of the external loading. Fig. 4(a, b) shows the (0002) pole figure obtained by EBSD measurements on the cross-sectional area of the undeformed TD and RD samples. The angular deviation of the loading axis (red circle) and the c-axis positions (black circle) with their highest texture intensity are ~90° and ~60° away for the TD and RD samples, respectively, reflecting the specific hot-rolled texture of α-Ti [4]. During bending, the inner region is subjected to compression, and the outer region experiences tensile loading. The possible maximum Schmid factor (SF) values of two twinning systems and their variation with the loading direction are given in Table 1.

Table 1: Possible maximum SF values of ET and CT and their variation with the loading direction.

Sample Nomenclature	Loading direction	Max. SF (Compression)		Max. SF (Tension)	
		$\{10\bar{1}2\}$ ET	$\{11\bar{2}2\}$ CT	$\{10\bar{1}2\}$ ET	$\{11\bar{2}2\}$ CT
TD bend	RD	0.49	0	0	0.45
RD bend	TD	0.29	0.12	0.12	0.41

Both the loading conditions (compression, Fig. 4a, b) favor a tensile strain along the c-axis, leading to the formation of ET in the ICR. However, the fraction of ET in the ICR of the TD bend is higher than the RD bend because $(10\bar{1}2)$ ET has a relatively high SF value of 0.49 (Table 1). The $\{11\bar{2}2\}$ CT has a maximum SF value of 0.12 for the ICR of the RD bend sample, indicating that it may still be active. In contrast, all the CT variants had a SF value of 0; thus, the formation of $\{11\bar{2}2\}$ CT is completely suppressed for the ICR of the TD bend sample.

Similarly, for TD bend under tensile loading along RD (Fig. 4a), the $\{10\bar{1}2\}$ ET exhibits SF values of 0 or negative, indicating no activation, while $\{11\bar{2}2\}$ CT showed a maximum SF of 0.45, indicating its chances of formation. Under tensile loading along TD for RD bend (Fig. 4b), both $\{10\bar{1}2\}$ ET and $\{11\bar{2}2\}$ CT were operative due to a maximum SF value of 0.12 and 0.41, respectively. Although the $\{11\bar{2}2\}$ CT exhibited a higher SF than the $\{10\bar{1}2\}$ ET for the OTR of the RD, the ET fraction was higher than that of CT (Fig. 3b(iii)), owing to its lower critical resolved shear stress.

Conclusions

The activation of specific twin systems across the different regions of the bend sample is mainly governed by the loading direction and initial texture of the material. $\{10\bar{1}2\}$ ETs are dominantly formed in the ICR of the TD bend sample, while the OTR exhibited only a few $\{11\bar{2}2\}$ CT. In the RD bend sample, both ICR and OTR showed the activation of $\{10\bar{1}2\}$ ET along with a few $\{11\bar{2}2\}$ CT, however, the twin fraction was higher in the ICR than the OTR, but still lower than that observed in the ICR of the TD bend sample.

Acknowledgement

The work is funded by the 'Scheme for Promotion of Academic and Research Collaboration (SPARC/2024-2025/ADVMA/P3518), Ministry of Education, Government of India' and 'SGDRI, SRIC, IIT Kharagpur, India'.

References

[1] W. Abd-Elaziem, M.A. Darwish, A. Hamada, W.M. Daoush, Titanium-Based alloys and composites for orthopedic implants applications: A comprehensive review, Mater Des 241 (2024). https://doi.org/10.1016/j.matdes.2024.112850.

[2] B. Jeon, M.S. Kim, S.H. Choi, Y. Jeong, Finite element analysis using elasto-visco-plastic self-consistent polycrystal model for E-form Mg sheet subjected to bending, J. Magnes. Alloys 11 (2023) 1393–1407. https://doi.org/10.1016/j.jma.2022.10.017.

[3] S.K. Nayak, J. K. Banikya, D.K. Chouhan, S. Biswas, Heterogeneous bending behavior of hot-rolled α-Titanium sheet via microstructure and texture evolution, Materialia (2026) 102672, https://doi.org/10.1016/j.mtla.2026.102672.

[4] J.W. Won, C.H. Park, S.G. Hong, C.S. Lee, Deformation anisotropy and associated mechanisms in rolling-textured high-purity titanium, J. Alloys Compd. 651 (2015) 245–254. https://doi.org/10.1016/j.jallcom.2015.08.075.

[5] M.H. Yoo Slip, twinning, and fracture in hexagonal close-packed metals. Metall Trans A 12, 409–418 (1981). https://doi.org/10.1007/BF02648537.

[6] D.K. Chouhan, A.K. Singh, S. Biswas, C. Mondal, On the strain-hardening behavior and twin-induced grain refinement of CP-Ti under ambient temperature compression, Metall. Mater. Trans. A 50 (2019) 2255–2268. https://doi.org/10.1007/s11661-019-05181-5.

[7] D.K. Chouhan, S. Mondal, B. Bhattacharya, S. Biswas, Role of slip and twinning on strain hardening and correlation with geometric hardening, latent hardening, and grain boundary strengthening in titanium, Int. J. Plast. 161 (2023) 103516. https://doi.org/10.1016/j.ijplas.2023.103516.

[8] S. Biswas, P.C. Gautam, A.J. Shukla, D.K. Chouhan, Dynamic recrystallization and its effect on microstructure and texture evolution in magnesium alloys, in: Encyclopedia of Smart Materials, Elsevier, 2021, pp. 476–481. https://doi.org/10.1016/B978-0-12-815732-9.00016-4.

Fickian diffusion and Flory-Rehner analysis of hydrocarbon swelling in stainless steel-integrated EPDM-UHMWPE hoses for AAR operations

Anjitha P S[1], Kshitija VAIDYA[1], Pratyush BAGARIA[2], Yash AVHAD[3], Balasubramanian KANDASUBRAMANIAN[1*]

[1]Department of Metallurgical and Materials Engineering, Defence Institute of AdvancedTechnology (DU), Giri Nagar, Pune - 411025, Maharashtra, India

[2]Department of Mechanical Engineering, MIT-World Peace University, Kothrud, Pune - 411038, India

[3]Department of Mechanical Engineering, SRM Institute of Science and Technology, Kattankulathur, Chennai - 603203, India

*meetkbs@gmail.com

Keywords: Multilayer composites, Aerospace refueling solutions, EPDM, UHMWPE, Thermo-mechanical simulations, Flory-Rehner Equation

Abstract. Air-to-air refueling (AAR) hose systems endure severe cyclic pressurization, dynamic bending, vibrational loading, and extended hydrocarbon exposure across -40 °C to 80 °C. Conventional fluorocarbon or nitrile butadiene rubber hoses reinforced with aramid fibers exhibit hydrocarbon diffusion, plasticizer migration, and interfacial delamination, compromising structural integrity and service durability. This investigation presents a novel multilayer composite incorporating ethylene propylene diene monomer (EPDM) elastomer for viscoelastic compliance and chemical resistance, austenitic SS-305 stainless steel mesh for tensile reinforcement and stress distribution, and ultra-high molecular weight polyethylene (UHMWPE) woven sheath as an effective diffusion barrier. Hot compression moulding (190 °C, 10 MPa, 60 min) with controlled cooling ensures robust interfacial bonding. Swelling tests (ASTM F146) in petrol demonstrated controlled 51% volumetric expansion within 7 hours, attaining equilibrium. Diffusion kinetics conformed to Fickian behaviour ($D \approx 0.033$-0.034 mm²/h). Flory-Rehner analysis yielded $\chi \approx 0.83$, confirming poor solvent affinity and constrained molecular mobility. The crystalline UHMWPE sheath suppressed EPDM chain relaxation, restricted ingress, and stabilized interfaces under thermal-solvent stress. Tribological evaluation (ASTM G99, 10 N, 263.9 m) revealed stable UHMWPE-derived transfer film formation, yielding friction coefficient 0.26-0.32 and specific wear rate 1.33×10^{-2} mm³ N^{-1} m^{-1}, evidencing superior durability and cohesion. The EPDM-UHMWPE-SS-305 composite delivers enhanced barrier performance, wear resistance, and delamination resistance, enabling reliable indigenous AAR hose solutions.

1. Introduction

Air-to-air refuelling (AAR) is a critical aerospace operation that enables in-flight fuel transfer from a tanker to a receiver aircraft, extending operational range and endurance without ground refuelling. The probe-and-drogue configuration predominates due to its structural simplicity, interoperability, and adaptability across diverse airframes [1]. This system utilises a flexible hose terminating in a conical drogue, into which the receiver's probe engages for sealed, pressure-regulated fuel delivery. Operational efficiency and safety depend on the hose assembly's mechanical resilience, fuel compatibility, and aerodynamic stability under cyclic tensile/torsional stresses, vibration-induced fatigue, and variable thermal, pressure, and aerodynamic loads [2].

The hose-conduit assembly serves as the central mechanical and fluidic pathway for high-pressure fuel transfer while withstanding aerodynamic drag, oscillatory bending, torsional vibration, and fluctuating pressures. It must maintain dimensional fidelity, sealing integrity, and laminar fuel flow under severe dynamic excitation [3]. Polymeric hose architectures, typically multilayered elastomeric and thermoplastic materials, demand negligible volumetric swelling during continuous aircraft fuel exposure, low hydrocarbon vapour permeation, and high resistance to abrasion, fatigue, and coupling wear. Conventional designs incorporating fluorocarbon rubber (FKM), nitrile butadiene rubber (NBR), or polytetrafluoroethylene (PTFE)-lined aramid-reinforced composites are constrained by aromatic fuel diffusion, plasticiser migration, and delamination under cyclic flexure [4].

To overcome these limitations, this study introduces a groundbreaking multilayer composite architecture that integrates stainless steel (SS) mesh as central reinforcement for superior tensile and fatigue strength, ethylene propylene diene monomer (EPDM) elastomer encapsulating the metallic core for viscoelastic compliance, chemical resistance, and interfacial sealing, and ultra-high molecular weight polyethylene (UHMWPE) fabric overlay for exceptional abrasion resistance, low surface energy, and dimensional stability. Fabricated via precision compression moulding, this hitherto unexplored synergistic structure achieves high interfacial adhesion, uniform consolidation, and enhanced chemo-mechanical resilience. By strategically reconciling minimal hydrocarbon-induced swelling with superior wear and fatigue endurance, the design establishes a pioneering framework for next-generation, delamination-resistant AAR refuelling hoses.

2. Materials and Methods

2.1 Materials and Fabrication of Composites

The assembled multilayer composite constituted of locally sourced SS305 stainless steel mesh sandwiched between 1mm EPDM (HEMRL, Pune), 1mm UHMWPE fabric (DEBEL, Bangalore) and lastly capped by 2mm EPDM at both ends. For fabrication, the multilayer assembly was subjected to compression moulding at 190°C under 10 MPa for 60 minutes, followed by room temperature cooling to minimize thermal stresses and enhance structural integrity.

2.2 Characterization

2.2.1 Diffusion and Swelling Characteristics

The diffusion behaviour of hydrocarbon fuel (petrol) in multilayer hose specimens was assessed via mass uptake measurements per ASTM F146. Fractional uptake ($M_t/M\infty$) was determined from periodic mass recordings until equilibrium was reached.Assuming one-dimensional through-thickness diffusion and plane-sheet geometry, the system was modelled using Fick's second law to estimate the apparent diffusion coefficient (D).In the early-stage regime ($M_t/M\infty < 0.4$), uptake follows the parabolic relation:

$$M_t/M\infty = \frac{4}{\sqrt{\pi}}\left(\frac{Dt}{l^2}\right)^{1/2} \tag{1}$$

where l is sample thickness. The slope S of the linear $M_t/M\infty\ vs.\sqrt{t}$ plot yields:

$$D = (\pi\, l^2\, S^2)/16 \tag{2}$$

In the late-stage regime ($M_t/M\infty > 0.6$), uptake is described by the exponential decay:

$$1 - M_t/M\infty = (8/\pi^2)\, exp(-\pi^2 Dt/l^2) \tag{3}$$

D is obtained from the slope of $ln(1 - M_t/M\infty)\ vs\ t$.

Agreement between D values from both regimes confirms Fickian diffusion and quantifies the composite's intrinsic barrier performance against hydrocarbon permeation under prolonged immersion [5], [6]. Thermodynamic polymer-fuel interaction was evaluated using Flory-Rehner theory on equilibrium swelling data to determine the polymer-solvent interaction parameter (χ):

$$\chi = \ln(1 - V_r) + V_r + \frac{\bar{V}_s}{V_r^2}(\frac{1}{2})(\frac{V_r^{1/3}}{V_{r,eq}} - \frac{V_r}{2}) \quad (4)$$

where V_r is the polymer volume fraction in the swollen state and $\bar{V}_s$ is the solvent molar volume. χ quantifies thermodynamic affinity, lower values indicate stronger fuel uptake and higher swelling propensity. The combined Fickian diffusion analysis and Flory-Rehner evaluation provide a comprehensive assessment of kinetic transport and molecular-level interaction mechanisms governing hydrocarbon sorption in the multilayer hose architecture [7].

2.2.2 Tribological Performance

The interfacial integrity of the multilayer hose composite was evaluated using a tribological pin-on-disc test in accordance with ASTM G99. The test was performed on a stainless-steel disc rotating at 300 rpm for 1200 seconds, employing composite specimens of 8 × 5 × 6 mm under a normal load of 10 N. The total sliding distance, calculated from the test parameters, was approximately 263.9 m[8]. The wear behaviour was quantitatively assessed using the Archard wear model in its experimental form, expressed as

$$k' = \frac{V}{F_N s} \quad (5)$$

where $k'$is the specific wear rate (mm^3/Nm), V the wear volume obtained from the measured wear depth and contact area, F_N the applied load, and s the total sliding distance. A low specific wear rate coupled with a stable friction coefficient is expected to confirm robust interfacial bonding, high wear endurance, and superior mechanical stability of the multilayer hose architecture under cyclic tribological conditions[9].

3. Results and Discussions

3.1 Swelling Behaviour

Swelling tests in petrol (ASTM F146) on the EPDM-UHMWPE-SS 305 multilayer composite showed rapid initial uptake: mass increased from 4.38 g to 6.63 g, diameter from 15 mm to 34 mm, and height from 6 mm to 10 mm over 7 hours, yielding ~51% volumetric expansion. Equilibrium was reached thereafter, with negligible changes up to 24 hours total immersion. This controlled profile contrasts with conventional single-layer EPDM or NBR, which often exhibit >100-200% uncontrolled swelling and prolonged non-equilibrium in aromatic hydrocarbons [10], [11]. The restrained swelling results from the hierarchical structure: the highly crystalline UHMWPE sheath physically confines EPDM chain mobility, while the impermeable SS 305 mesh disrupts diffusion pathways, limiting free-volume expansion and solvent-induced plasticisation. Compared to oil-extended EPDM compounds, where elevated plasticiser content reduces crosslink density and increases swell ratios [11], the barrier layers here provide superior long-term dimensional stability. This ensures fidelity under cyclic pressurisation and thermal cycling (-40 °C to 80 °C), reducing risks of kinking or sealing failure in probe-and-drogue AAR systems and supporting indigenous aerospace hose development amid supply disruptions.

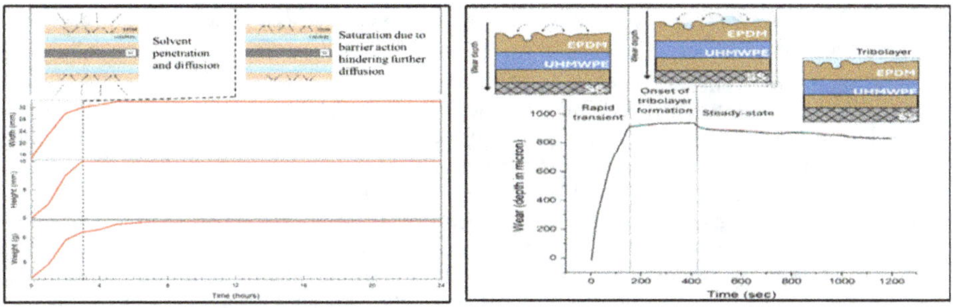

Fig 1. (a) Swelling Behaviour in Petrol *(b)Tribological Analysis of Composite*

3.2 Tribological Performance

Pin-on-disc testing (ASTM G99, 10 N load, 300 rpm, ≈263.9 m sliding) of the EPDM-UHMWPE-SS 305 multilayer composite against stainless steel counterface revealed three characteristic wear regimes. The initial transient regime (0-200 s) showed rapid wear depth increase from EPDM asperity conformance and microploughing, generating debris and high interfacial shear. In the tribolayer formation phase (200-400 s), shear-induced smearing of the UHMWPE sublayer produced a thin, coherent transfer film on the steel disc, enabled by UHMWPE's self-lubricating interlamellar shear mechanism that reduced asperity contact and stabilized friction. Beyond 400 s, steady-state regime emerged with asymptotic wear stabilization, as the protective transfer film supported load and shielded the SS 305 mesh from abrasion. Final wear depth ≈0.88 mm, wear volume 35.2 mm^3 (8 × 5 mm^2 contact area), and Archard specific wear rate $k = 1.33 \times 10^{-2}$ $mm^3/N{\cdot}m$ confirmed controlled adhesive-abrasive wear. The coefficient of friction rose from ≈0.2 to ≈0.32 during run-in, then stabilized at 0.26-0.32 in steady-state with minimal fluctuation-consistent with UHMWPE-steel dry sliding literature. Synergistic contributions from UHMWPE transfer film, EPDM viscoelastic damping, and SS mesh stress redistribution suppressed severe ploughing and fatigue cracking, projecting outer sheath integrity >1,000 cycles for aerospace refuelling hoses under cyclic drogue-probe coupling and vibrational loading.

3.3 Fickian Diffusion Analysis

Diffusion kinetics of hydrocarbons in the EPDM-UHMWPE-SS 305 multilayer composite followed pure Fickian behaviour, confirmed by close agreement between early- and late-stage diffusion coefficients. Early-stage analysis ($M_t/M_\infty < 0.4$), using the parabolic relation Eq.1 yielded D = 0.034 mm^2/h. Late-stage analysis ($M_t/M_\infty > 0.6$), based on the exponential decay Eq. 3 , gave D = 0.033 mm^2/h. The consistency (D ≈ 0.033-0.034 mm^2/h) excludes anomalous (Case II or relaxation-dominated) transport, which is frequently reported in highly filled or oil-extended EPDM under elevated temperature or humidity conditions [2], [12]. Literature diffusion coefficients for unfilled EPDM in toluene or analogous hydrocarbons typically fall in the range 0.036-0.36 mm^2/h (converted from 10^{-6}-10^{-5} cm^2/s) , highlighting the substantial reduction achieved here through synergistic barrier reinforcement. The low D values stem from the high crystallinity and low free volume of the UHMWPE sheath, which creates tortuous diffusion paths, combined with the SS 305 mesh's contribution to morphological stabilisation and stress redistribution. This architecture effectively suppresses long-term vapour permeation, a primary degradation mechanism in conventional FKM- or NBR-lined hoses [13] and outperforms hierarchical porous systems where non-Fickian effects prevail at higher filler loadings . The purely Fickian regime ensures minimal fuel loss and sustained structural integrity during prolonged exposure, enhancing operational reliability under vibrational loading and thermal cycling (-40 °C to 80 °C). These characteristics position the composite as an advanced solution for next-generation AAR hose conduits [1].

3.4 Flory-Rehner Thermodynamic Analysis

Thermodynamic polymer-solvent interactions were evaluated using Flory-Rehner theory on equilibrium swelling data, yielding a polymer-solvent interaction parameter $\chi \approx 0.83$ (mass/additivity method). Values of $\chi > 0.5$ indicate petrol as a thermodynamically poor solvent for the EPDM matrix, reflecting unfavourable mixing enthalpy and limited affinity between non-polar hydrocarbon chains and the crosslinked network [12], [14]. This elevated χ accounts for the restricted 51% swelling and rapid equilibrium attainment, as weak osmotic forces suppress chain plasticisation and network expansion. In filled EPDM systems, filler type and loading modulate χ via changes in crosslink density [2]; here, the UHMWPE sheath's confinement effect enhances this, limiting excessive free-volume formation under solvent stress and outperforming conventional elastomers (e.g., NBR with $\chi < 0.5$ in fuels, prone to delamination) [10], [13]. Comparative studies on HNBR/EPDM blends show analogous χ dependence on solvent molar volume and temperature ,confirming the multilayer design's optimised thermodynamic resistance. For AAR hoses, high χ minimises softening, oxidative ageing, and interfacial failure during prolonged fuel exposure, preserving sealing performance and fatigue life under complex stresses. This surpasses traditional architectures and supports robust, indigenously viable materials for strategic aerospace applications[4], [14].

4. Conclusion

This study successfully developed and characterized a novel multilayer composite hose integrating EPDM, UHMWPE fabric, and SS 305 mesh through hot compression moulding, targeting advanced air-to-air refuelling (AAR) applications. The synergistic architecture effectively overcomes key shortcomings of conventional fluorocarbon- or nitrile rubber-based hoses, including excessive hydrocarbon swelling, diffusion-driven plasticisation, interfacial delamination, and insufficient wear resistance under harsh thermo-mechanical cycling. Swelling experiments in petrol exhibited controlled 51% volumetric expansion, reaching equilibrium within 7 hours, owing to the barrier action of the crystalline UHMWPE sheath and impermeable SS mesh, which constrain EPDM chain mobility and limit free-volume expansion. Diffusion followed strict Fickian kinetics with low diffusion coefficients ($D \approx 0.033\text{-}0.034\ mm^2/h$), confirming highly obstructed hydrocarbon transport pathways. Flory-Rehner analysis revealed a polymer-solvent interaction parameter $\chi \approx 0.83$ (> 0.5), indicating poor solvent affinity and weak osmotic forces that minimise long-term plasticisation and dimensional instability across the operational temperature range of -40 °C to 80 °C. Tribological testing identified three wear regimes culminating in steady-state behaviour, enabled by a self-lubricating UHMWPE-derived transfer film; the specific wear rate of $1.33 \times 10^{-2}\ mm^3/N{\cdot}m$ and stable friction coefficient (0.26-0.32) demonstrate outstanding abrasion resistance and interfacial cohesion, safeguarding the SS reinforcement during repeated drogue-probe engagements. Collectively, the EPDM-UHMWPE-SS 305 composite provides a pioneering, delamination-resistant platform with superior chemical inertness, permeation barrier performance, and mechanical durability, making it well-suited for indigenous development of reliable AAR hoses amid geopolitical supply chain disruptions. Future investigations should focus on scaled prototypes under dynamic flight conditions to confirm real-world probe-and-drogue mission performance.

Acknowledgments

The authors would like to thank Dr. BHVS Narayana Murthy, Vice Chancellor, DIAT (DU) Pune, for constant encouragement. The authors would like to acknowledge Ms. Shruti Gupta, Ms. Neelaambhigai Mayilswamy, Ms. Manisha Singh, and Ms. Payal Varma for their guidance and technical support.

References

[1] P. R. Thomas, U. Bhandari, S. Bullock, T. S. Richardson, and J. L. Du Bois, "Advances in air to air refuelling," Prog. Aerosp. Sci., vol. 71, pp. 14–35, Nov. 2014. https://doi.org/10.1016/j.paerosci.2014.07.001

[2] J. Ren and Q. Quan, "Progress in modeling and control of probe-and-drogue autonomous aerial refueling," Chin. J. Aeronaut., vol. 37, no. 5, pp. 6–26, May 2024. https://doi.org/10.1016/j.cja.2023.11.011

[3] N. Fezans and T. Jann, "Towards automation of aerial refuelling manoeuvres with the probe-and-drogue system: modelling and simulation," Transp. Res. Procedia, vol. 29, pp. 116–134, 2018. https://doi.org/10.1016/j.trpro.2018.02.011

[4] J. Ament, J. Lachmann, J. Schmelz, and L. Wandrey, "Design and Assessment of Fighter Pilot Assistance Systems for Air-to-Air Refuelling with Probe-and-Drogue-Equipment," CEAS Aeronaut. J., vol. 15, no. 4, pp. 1091–1110, Oct. 2024. https://doi.org/10.1007/s13272-024-00756-4

[5] C. Darko, "The link between swelling ratios and physical properties of EPDM rubber compound having different oil amounts," J. Polym. Res., vol. 29, no. 8, p. 325, Aug. 2022. https://doi.org/10.1007/s10965-022-03179-z

[6] X. Jiang, X. Yuan, X. Guo, F. Zeng, H. Wang, and G. Liu, "Study on the application of Flory–Huggins interaction parameters in swelling behavior and crosslink density of HNBR/EPDM blend," Fluid Phase Equilibria, vol. 563, p. 113589, Jan. 2023. https://doi.org/10.1016/j.fluid.2022.113589

[7] P. Fan, "A universal expression for chemical potential in vulcanized-rubber swelling and permeation," J. Elastomers Plast., p. 00952443251414348, Dec. 2025. https://doi.org/10.1177/00952443251414348

[8] S. Ragunath, N. Radhika, S. Mahesh, and T. Ramkumar, "Improving mechanical and tribological performance of TiB reinforced Cu composites through microwave sintering," Results Eng., vol. 29, p. 108780, Mar. 2026. https://doi.org/10.1016/j.rineng.2025.108780

[9] D. Singh, R. R. Kumar, P. K. Karsh, M. K. Kumar, G. S. Mandal, and H. M. S. Ali, "Investigation on tribological and mechanical behaviour of GFRP composites with varying weight percentages of nano-graphite powders," Sci. Rep., vol. 15, no. 1, p. 34924, Oct. 2025. https://doi.org/10.1038/s41598-025-18690-8

[10] J. L. Graham, R. C. Striebich, K. J. Myers, D. K. Minus, and W. E. Harrison, "Swelling of Nitrile Rubber by Selected Aromatics Blended in a Synthetic Jet Fuel," Energy Fuels, vol. 20, no. 2, pp. 759–765, Mar. 2006. https://doi.org/10.1021/ef050191x

[11] P. Kulasekaran, B. Maria Mahimai, G. Sivasubramanian, H. Pushparaj, and P. Deivanayagam, "Zinc-trimesic acid metal–organic framework incorporated sulfonated poly(ether ether sulfone) based polymer composite membranes for fuel cell," Polym. Eng. Sci., vol. 62, no. 9, pp. 2986–2997, Sep. 2022. https://doi.org/10.1002/pen.26079

[12] S. Akhlaghi, A. M. Pourrahimi, C. Sjöstedt, M. Bellander, M. S. Hedenqvist, and U. W. Gedde, "Degradation of fluoroelastomers in rapeseed biodiesel at different oxygen concentrations," Polym. Degrad. Stab., vol. 136, pp. 10–19, Feb. 2017. https://doi.org/10.1016/j.polymdegradstab.2016.12.006

[13]R. J. Pazur and J. G. Cormier, "COMPATIBILITY STUDY OF LOW-TEMPERATURE–CAPABLE FLUOROELASTOMERS IN JET FUELS," Rubber Chem. Technol., vol. 88, no. 4, pp. 660–675, Dec. 2015. https://doi.org/10.5254/rct.15.84873

[14]G. Butuc, K. V. Leerdam, B. Rossenaar, A. Talma, and A. Blume, "Deciphering the crosslink mechanism of dual cure EP(D)M and CTS rubber compounds for reduced oil swell," Polym. Test., vol. 145, p. 108736, Apr. 2025. https://doi.org/10.1016/j.polymertesting.2025.108736

Keyword Index

About the Editors

Dr. Syed Naismul Alam is an Associate Professor in the Department of Metallurgical and Materials Engineering, National Institute of Technology Rourkela, Rourkela, India. He did his Bachelor of Technology (B.Tech.) in Metallurgical and Materials Engineering Department at Indian Institute of Technology Kharagpur, India in 1997. He did his Masters in Materials Science and Engineering at University of Massachusetts Lowell, USA in 2001. He did his PhD in Metallurgical and Materials Engineering Department at Indian Institute of Technology Kharagpur, India in 2010 in the area of "Tungsten-based nanocomposites developed by powder metallurgy (PM) route". His research area includes mechanical alloying (MA), nanocomposites, 2D materials and graphene. He has published about 40 research papers in highly reputed international journals and has a patent granted in his area of research and has also filed two more patents in this area. He is the author of 3 books on MATLAB and Simulink and has also supervised 3 PhD students and several graduate and undergraduate level students. Preently his research area includes 2D materials, graphene, mechanical alloying, nanocomposites and intermetallics.

Dr. Krishna Dutta is working as associate professor in the Department of Metallurgical and Materials Engineering at the National Institute of Technology Rourkela (NITR), Odisha, India. He earned his Bachelor of Engineering in Metallurgical Engineering from Jadavpur University, Master of Technology from IIT Kanpur, and completed his Ph.D. in Metallurgical and Materials Engineering at IIT Kharagpur (2012). At NIT Rourkela, Dr. Dutta teaches the courses related to mechanical testing, fracture mechanics and supervises research students in advanced materials topics. His research interests focus specifically on fatigue and fracture behavior of materials, with contributions to the understanding of mechanical properties and performance of 3D-printed materials, high entropy alloys. He has authored over 50 international journal articles, several conference papers, and book chapters. Dr. Dutta is involved in sponsored and consultancy projects, and has guided multiple doctoral students. In addition to his research and teaching roles, he also serves in academic leadership capacities, including as Professor In-Charge of the Accreditation and Ranking Cell at NIT Rourkela.

Dr. Kumud Kant Mehta is an Associate Professor in the Department of Metallurgical and Materials Engineering at the National Institute of Technology (NIT) Rourkela, with over 22 years of extensive experience encompassing academia, defence research, and the steel industry. His core expertise lies in the domain of mechanical metallurgy, with a strong emphasis on the development, processing, characterization, and performance evaluation of advanced engineering and aerospace alloys, including nickel-, titanium-, aluminum-based alloys, and special steels. Prior to his academic career, Dr. Mehta served in the Directorate General of Aeronautical Quality Assurance (DGAQA), Ministry of Defence, Government of India, where he held key scientific positions. During his tenure, he was actively involved in the indigenization, development, and quality assurance of critical aeronautical materials and components, including those used in aero-engines such as the AL31FP (Sukhoi) engine. His contributions span the entire lifecycle of materials—from raw material qualification to final assembly, testing, and failure investigation—along with participation in aircraft accident and incident analysis. He has played a significant role in the development and certification of strategic materials such as Ni-based superalloys, titanium alloys, maraging steels, and aerospace-grade aluminum alloys.

Dr. Mehta obtained his Ph.D. in Metallurgical Engineering from IIT (BHU), Varanasi, where his research focused on texture and mechanical properties of nickel-based alloys. He completed his M.Tech. from IIT Kharagpur as a DRDO-sponsored candidate, securing the top rank in his batch, and earned his B.E. in Metallurgical Engineering with distinction from BIT Sindri. His research interests include microstructure–texture–property correlations, thermo-mechanical processing, anisotropy in mechanical behavior, constitutive modeling (including VPSC), creep and fatigue behavior, fracture mechanisms, and environmental degradation of materials. He has published a substantial number of research papers in reputed international journals and conference proceedings, and has contributed to a book on aerospace materials and technologies. He is actively involved in sponsored research projects funded by national agencies, consultancy assignments, and the supervision of postgraduate and doctoral students in advanced materials research. He has also contributed to the preparation of material specifications and quality plans for defence applications. Dr. Mehta is a recipient of several prestigious recognitions, including the DRDO Agni Award for self-reliance in materials indigenization, along with multiple appreciation awards for his contributions to strategic aerospace programs. He is a life member of several professional bodies, including the Indian Institute of Metals and the Indian Society for Non-Destructive Testing, and continues to contribute to academic, research, and professional activities in the field of materials engineering.

www.ingramcontent.com/pod-product-compliance
Lightning Source LLC
LaVergne TN
LVHW022304200926
848481LV00004B/1210
9781644904121